*Joanne —
Be close to the sea.
Live well and enjoy!
Jill Grover Olla*

⋙ ECOLOGICAL ⋘
FOOD *for* THOUGHT
on SEAFOOD

Jill J. Grover

All rights reserved. No part of this book shall be reproduced or transmitted in any form or by any means, electronic, mechanical, magnetic, photographic including photocopying, recording or by any information storage and retrieval system, without prior written permission of the publisher. No patent liability is assumed with respect to the use of the information contained herein. Although every precaution has been taken in the preparation of this book, the publisher and author assume no responsibility for errors or omissions. Neither is any liability assumed for damages resulting from the use of the information contained herein.

Cover photograph by author.

Copyright © 2013 by Jill J. Grover

ISBN 978-0-7414-9799-4 Paperback
ISBN 978-0-7414-9800-7 eBook
Library of Congress Control Number: 2013913557

Printed in the United States of America

Published August 2013

INFINITY PUBLISHING
1094 New DeHaven Street, Suite 100
West Conshohocken, PA 19428-2713
Toll-free (877) BUY BOOK
Local Phone (610) 941-9999
Fax (610) 941-9959
Info@buybooksontheweb.com
www.buybooksontheweb.com

Table of Contents

Introduction v
What draws us to the sea? 1
What is our relationship with Earth's oceans? 3
What are the benefits of including seafood in a diet? 12
What are the health benefits of omega-3 fish oils? 18
How are heavy metals biomagnified through the food chain? 23
    Other persistent contaminants 28
Do the benefits of eating seafood during pregnancy exceed the risks? 32
What should we know about naturally occurring toxins in seafood? 35
    Shellfish poisoning 37
    Fish poisoning 43
Is any seafood really safe to eat? 49
What's the truth behind shellfish myths? 51
Is a seafood diet good for the planet? 55
How has fishing evolved? 57
What makes a fishery sustainable? 63
Which fisheries are sustainable? 66
Which fisheries are not sustainable? 74
What is bycatch and why is it important? 85
What are the ecological impacts of capture fisheries? 91
What about aquaculture? 97
How does aquaculture impact wild fish populations? 101
    Wild feed 101
    Wild seed 104
    Partly farmed, partly wild 106
Which is better, farm-raised or wild shrimp? 109
Is farmed salmon as good as, or good for wild salmon? 111

What about open-sea fish farming? ... 115
    Cobia ... 116
    Moi ... 116
    Hawaiian yellowtail ... 117
    Mussels ... 119
Is any aquaculture sustainable? ... 120
Can we trust imported seafood? ... 123
Will world demand for seafood exceed world seafood production? ... 127
How can we make better use of our seafood resources? ... 130
Can we switch to underutilized species? ... 134
Can we utilize fish more efficiently? ... 137
    Surimi ... 138
    The indigenous approach ... 142
Sustainable Seafood Recipes ... 145
    Crab cakes ... 145
    Crab and chanterelle couscous ... 147
    Planked salmon ... 148
    Salmon en papillote (Salmon in parchment) ... 150
    Salmon loaf ... 152
    Penne with albacore belly meat ... 153
    Salmon head risotto ... 155
    Smoked oyster risotto ... 159
    Smoked albacore pizza ... 160
    Smoked salmon enchiladas ... 162
    Mussels fra diavolo ... 164
    Spicy linguine with mussels in a red pepper and bruschetta sauce ... 166
    Pesto pink shrimp on spaghetti ... 167
End notes ... 169
Index ... 185
Acknowledgements ... 189

Introduction

Recently, a longtime friend asked me a seemingly simple question. She wanted to know which seafoods were good for her to eat. As a dedicated consumer of fish and shellfish, I answered her largely in terms of goodness from a human health perspective. However, soon after she posed that question, I began thinking about goodness from a larger point of view, in terms of a global health perspective.

Having spent my career as a marine feeding ecologist, I felt haunted by all the answers that I could have provided. All sorts of ecological issues flooded my mind. The marine ecologist in me needed to think bigger. Over the last ten years, I have read thousands of pages of books and scientific articles on marine topics, and hope to distill some of that science into terms that are easily understandable to a general audience. I live close to the sea, in Oregon, and eat close to the sea.

My goal here is to answer the larger question of ecological goodness. In the sense that we are what we eat, gaining knowledge about the food that we eat is gaining self knowledge and identifying our trophic position in the universe. Beyond that, I hope to encourage the enjoyment of globally sustainable seafood from functional, diverse ecosystems by sharing a few recipes.

Now is a very confusing time to be trying to figure out how healthful various seafood choices are. Over the last decade, new claims or warnings about the healthfulness or safety of seafood products have frequently been broadcast across the full spectrum of news media sources. These warnings have been issued by federal and state government agencies, as well as by independent groups. They are hard to ignore, and sometimes even harder to understand, when seemingly conflicting messages are released by the same group at different times, or by different groups at the same time.

Even more confusing is when seafood products are mislabeled or misrepresented in order to appear more healthful than they are. The flesh of some species is quite similar in appearance to other species. Thus, unscrupulous vendors can substitute inferior, cheaper fish species for more desirable and expensive species. Farmed-raised salmon has been sold at wild-caught prices. Shark steaks have been sold as swordfish. Imitation crab has been sold as true crab meat. Now is a great time to be a well-informed seafood consumer.

As man-made contaminants are showing up in the tissues of top marine predators that live far away from continents and feed deep in the ocean, clearly, planet Earth is one big inter-connected system. What we do in one place has consequences that are felt throughout our global village. We need to act responsibly towards the oceans before we destroy what remains of their bounty.

What draws us to the sea?

Humans love the sea, and much of what it represents. We have a strong connection with the sea, and have long been fascinated by sea stories. For centuries, the sea has featured prominently in classic literature. True ocean adventure tales comprise a popular genre of literature, encompassing voyages, disasters, and searches for ship wrecks. Poets have also answered the call of the sea. Sea stories have long been a part of children's literature as well, including fairy tales. Wet and wild tales from the sea have long entertained us and provided an escape from the earthly, dry mundaneness of our everyday lives.

Over the last century, a tremendous body of scientific literature has been published that reports on and interprets a broad spectrum of biological, chemical, and geo-physical data that have been collected in ocean environments. Unfortunately, the bulk of scientific literature is neither readily available, nor easily understandable to the general public. However, through the waning decades of the last millennium and continuing into the new, an increasing number of scientists and authors have written accessible ocean literature, books about the ocean and marine issues using terms a layperson can understand, and get excited about.

Marine themes have permeated music as well, including both classic and popular pieces. Sea shanties comprise an entire genre of music. Although they are now considered folk songs, shanties functioned as working songs that provided cadence to a crew of sailors hauling lines, or performing other chores on sailing vessels. In the early 1960s, surfer music became a popular genre. Surf music has largely faded into obscurity, but marine themes still occasionally pop up in new music.

The earliest examples of marine art (portraying vessels and water) date back to a time before written language. These primitive images were petroglyphs, engraved on rock. Although very old images offer extraordinary glimpses into history, the compelling nature of marine art only began following significant advancement and cultural refinement of human society. Classic sea paintings may be pleasing and quite familiar to many of us, but the visual images of the sea that stick with us longest may be scenes from popular movies or television series.

Many of us live near the sea, and vacation on or near the sea, coast, or beach. While taking long walks on the seashore, we comb the beaches searching for sea shells, glass floats, polished stones, driftwood, and other treasures. We participate in a myriad of recreational activities on or in the sea (including swimming, surfing, snorkeling, scuba diving, boating, sailing, fishing, whale watching, bird watching, kite flying, kite boarding, wind surfing, and beach volleyball). We build ephemeral sand castles and driftwood installations on the beach. We select seafood as an important part of our diet. Seascapes appeal to us. Watercolors appeal to us. Water sounds provide the soothing noise that lulls us to sleep: rain drops, streams, waterfalls, as well as the lapping of ocean waves. For those of us who live in cities, white-noise machines now mimic these great sounds of nature.

Biologically, we are all connected to the sea, in as much as life on earth originated in the sea. An abundance of fossil evidence indicates that all terrestrial animals evolved from ancestors in the sea. And seas continue to play a vital role in sustaining life on land. The sea represents a last frontier to many of us, an often unpredictable, somewhat mysterious, powerful place. Yet its high energy seduces us to go with its ebbs and flows.

What is our relationship with Earth's oceans?

Seawater covers more than two thirds of our planet. As a result, oceans profoundly affect Earth's atmosphere and climate. Solar radiation warms surface waters of the oceans, and releases water vapor into the atmosphere. Water that evaporates from the surface of the ocean forms clouds and drives the water cycle (also known as the hydrologic cycle) that redistributes water around the globe, and provides our most abundant source of the rainfall that sustains life on Earth. Most of the oxygen in our atmosphere was produced by phytoplankton (microscopic plants, or micro-algae, and their ecological analogues) in the ocean, through photosynthesis, a process which, in the presence of sunlight and chlorophyll, converts the raw materials carbon dioxide and water, into carbohydrates and oxygen. In the course of photosynthesis, marine plants (as well as terrestrial plants) perform two extremely valuable atmospheric functions that benefit animal life: they increase the supply of oxygen and they decrease carbon dioxide.

At the same time, global climate is greatly influenced by ocean conditions. Global ocean currents moderate the climate in distant lands. For example, the Gulf Stream, a fast, intense current along the western border of the North Atlantic Ocean, warms the climate in Britain and Europe with water that flows out of the Gulf of Mexico. However, as global warming proceeds, some of the big hydrological "engines" that drive the Gulf Stream current, such as the sinking of super-cooled water in the Greenland Sea, could be slowing down.[1,2] Previously, many columns of dense, cold water sank to the seabed at great depths. This sinking, or down-welling, of cold water acted as an engine that pulled warm surface water from the south, and continued the circulation of the warm water. Oceanographic observations made in 2004, indicated that these cold-water columns may be disappearing. Any slowdown in the Gulf Stream would mean less heat would reach Europe and could result in sharp temperature declines in Britain and northwestern Europe. However, in 2006, a different analysis suggested that the earlier observations were merely examples of year-to-year variability in circulation patterns in the Atlantic Ocean,[3] a system known for high variability, and suggested even if this conveyor of warm water to high latitudes were slowing, decades would pass before any changes would be noticeable.

Another oceanic circulation pattern, the El Niño, has clearly been shown to produce profound climatic effects in very distant geographical zones. Very briefly, El Niño conditions occur when weakened trade winds in the southwestern Pacific Ocean, near Indonesia, allow warmer water to flow east. As a result, warm surface water builds up off South America, and the water temperature increases at depth as well. Unusually strong El Niño events can produce very severe, cascading and sometimes confounding impacts on both oceanography and climate, over a broad geographical area, ranging well up the coast of North America. During El Niño events, normal patterns of upwelling (the rising to the surface) of cold, nutrient-rich, bottom water are interrupted, and without nutrients, marine food chains may collapse. The collapse of food chains negatively impacts fisheries and the seabirds and mammals that feed on them, as well as all the people who depend on the marine food chain for food security or to earn a livelihood. The name El Niño is historically derived from the timing of the onset of anomalous oceanographic conditions off South America, near Christmas time. In Spanish, "El Niño" signifies the Christ child.

Throughout the world, oceans moderate coastal climates, due to the heat capacity of water. Heat capacity, also known as specific heat, measures how much heat energy is required to raise the temperature of a one gram of a chemical substance one degree Celsius. Oceans store tremendous amounts of heat from solar radiation, and generally take quite a long time to warm or cool significantly. As a result, along coastal regions of continents, ambient temperatures are warmer in winter and cooler in summer compared to inland climates. Because land heats and cools faster than water, air temperatures at the seashore tend to rise slower in spring, and drop slower in the fall, compared with air temperatures farther inland. Continental climates, at locations that are distant from sea coasts, experience harsher temperature extremes in both winter and summer than coastal climates.

One of the most visible links between ocean conditions and weather has been the intensity of the recent Atlantic hurricane seasons. Hurricanes originate in the tropics. They start as thunderstorms over warm (>80°F/27°C) seas, under humid conditions, with convergent winds. As they progress from tropical depressions and storms into hurricanes, these weather systems derive their strength from high ocean temperatures. In 2005, warm, moist air over extremely warm water in the Gulf of Mexico provided the energy necessary to produce monster hurricanes such as Katrina and Rita. While the term "hurricane" is limited in its geographical use, to the North Atlantic Ocean, the Northeast Pacific Ocean – east of the dateline, and South Pacific Ocean east of 160°E, this type of storm occurs in a much wider area. In other parts of the world, similar ocean-generated tropical storms are known as typhoons in the Northwest Pacific Ocean, or as severe tropical cyclones in the Southwest Pacific Ocean and Indian Ocean.

Oceans shape our climate and atmosphere, and the presence of water on Earth has made our planet habitable. Long before life was present on land, fossil records have shown that life was present in Earth's oceans. While some of the exact mechanics of evolution may be legitimately debated, scientists accept the occurrence of evolution as a fact. In the millenia since the first animals crawled out of the sea to begin a land-based existence, adaptive radiation has produced a tremendous diversity of species. All life on Earth is still connected to the sea.

Coastal marine ecosystems, such as coral reefs, mangrove forests, kelp beds, and coastal wetlands, can protect shorelines from wave-generated erosion, even during tsunamis and hurricane storm surges. Wetlands and mangroves trap sediments. In a healthy, balanced ecosystem, microbes in those sediments can tie up, as well as detoxify, many of the pollutants that flow through them.

Many highly productive, spatially complex, coastal habitats, including wetlands, estuaries, salt marshes, seagrass beds, coral reefs, mangrove forests, kelp forests, and oyster beds, perform important ecological roles in the early life history of many marine species. These ecologically rich environments function as essential nursery habitats for a majority of commercially and ecologically important marine fish species. Prime nursery grounds facilitate the growth and development of larval and early-juvenile fishes while offering spatial refuge from voracious predators. They provide both food and shelter for vulnerable life stages.

Despite their ecological importance within marine and coastal ecosystems, many coastal habitats are declining or being negatively impacted by a variety of human activities. Coastal development degrades habitats both structurally (e.g. through the draining and reclamation of wetlands and the removal of mangroves) and chemically (through nutrient enrichment, and pollution). In addition, coastal habitats are damaged by dredging, beach replenishment, non-sustainable fishing practices such as bottom trawling, and the introduction of invasive, non-native species through ballast water discharges and aquaculture. Naturally productive mangrove forests are being cut down and replaced with shrimp farms throughout the tropics and subtropics where they occur. In the name of progress and opportunity, coastal development has destroyed critical nursery habitats for many marine species.

Over the years, we have simultaneously treated Earth's oceans as our food basket and our playground; a pathway to adventure, exploration, and discovery; a giant conveyor belt for cheap labor, raw materials and finished products; a largely untapped supply of mineral, biological, and biochemical resources to be extracted; a war site, a dump site, and a septic tank. The oceans have been a protective moat around island nations, and kept some cultures apart, while bringing others together. For centuries, the seas have done an admirable job of living up to our expectations. However, as the human population on Earth rapidly increases, this delicate balance is changing.

Historically, our oceans have been viewed as a virtually limitless source of protein, capable of feeding the world's burgeoning population. As recently as the middle of the twentieth century, leading scientists suggested that the ocean's bounty could be harvested more intelligently, and when aided by focused research and technology, fisheries could provide abundant seafood to feed the masses.

Earth's oceans comprise the largest natural food resource on the planet. However, despite folklore to the contrary, ample evidence exists that these resources are finite and some fisheries have been overfished for hundreds of years. Historically, in seaport after seaport, when local stocks of preferred species were depleted, fishing boats ranged farther from home. Eventually, as new fishing grounds were depleted, the boats extended their range even farther. The long-distance, sailing fleet of Portuguese cod fishermen may have discovered the fertile fishing grounds of the New World long before Columbus sailed in 1492, but chose to remain quiet about their discovery rather than reveal their bountiful source of cod.[4] Legendary fishing banks became fishing targets of many nations, and not surprisingly, yields dropped. Today, the North Atlantic Ocean cod fishery is a classic example of a collapsed fishery.

National and international agencies have aimed to regulate the intense amount of fishing in productive areas like Georges Bank. Yet despite treaty obligations and biological data to the contrary, economic pressures kept the fisheries open until they collapsed, due to prolonged overfishing.

Unfortunately, the oceans have also been viewed by some as a dumping ground for all kinds of noxious materials. They were seen, in chemical terms, as giant buffer solutions. Quite simply, our oceans were treated as if they intrinsically possessed both the chemical capability and physical capacity to mitigate, neutralize, and/or dilute an amazing array of chemical compounds. This may have been true for small quantities of rather simple waste products. However, as human populations have grown, so have the volume, complexity, and toxicity of their waste products. Up until the tail end of the twentieth century, large amounts of chemical wastes were routinely dumped into the ocean.

Ocean dumping was seen essentially as an easy way to make pollutants disappear through dilution, dispersion, sedimentation, and degradation.[5] However, many of the man-made chemicals that have been dumped into the oceans are virtually indestructible, and cannot easily be destroyed. For example, from the 1940s to the 1970s, tons of recalcitrant chemicals including DDT (dichlorodiphenyl trichloroethane), an organic insecticide, and PCBs (polychlorinated biphenyls), a class of organic chemicals used in transformers, hydraulic fluids and paints, were dumped offshore of Los Angeles, fouling the marine environment near the Channel Islands. As late as the 1990s, radioactive wastes from nuclear reprocessing facilities in France, England and Scotland, were dumped into marine environments. Russian nuclear wastes were also routinely dumped into the Barents and Kara Seas. The US still allows ocean dumping of some low-level nuclear wastes. Recently, as a result of the tsunami in Japan (March 11, 2011), nuclear wastes contaminated the marine ecosystem offshore of the accident site. Beyond measuring radioactive isotope levels, the full marine impact of Fukushima's tsunami-induced nuclear accident will be hard to pin down exactly, as the marine environment was subjected to confounding physical and chemical insults from the earthquake, tsunami, toxicants, and nuclear accident.

From the 1920s into the 1990s sewage sludge and garbage were routinely barged off the coast of New York and New Jersey and dumped at sea, resulting in a seasonal dead zone on the sea bottom in the New York Bight. Dead zones are created by excessive nutrient enrichment, which over-stimulates algal growth. When massive algae blooms die off, they sink to the bottom, and decompose. The microbial break-down of dead algae consumes virtually all of the oxygen in the water, and creates a low-oxygen (hypoxic) environment that is hostile to most animal life. Dead zones vary in frequency and duration, ranging from episodic to seasonal to persistent. Although ocean dumping of sewage sludge and industrial waste was banned in the US, by the Ocean Dumping Ban Act of 1988, which became effective December 1991, dead zones are spreading around the world.[6] Many industries, such as paper mills and sewage treatment plants, continue to pipe "treated" effluent into the ocean directly, or indirectly through river drainage. While new permits for effluent discharge are often difficult to obtain, long-term effluent dumping is generally re-permitted with minimal public comment, due to grandfather clauses that often protect jobs more than the environment. In the best case scenario, only sewage that has undergone tertiary treatment is discharged into watersheds. However, in the worst case scenario, for example under conditions of torrential rain, storm surge, and/or flooding, waste-water treatment plants are unable to keep up with the extra demands on their pumps, and raw sewage can end up being discharged.

A massive, persistent, dead zone has developed in the northern Gulf of Mexico, downstream of the Mississippi River delta off the coast of Louisiana and Texas, as a result of years of excessive nutrient enrichment, particularly nitrogen from agricultural runoff. This hypoxic zone has been monitored annually since 1985. In 2002, the Mississippi-spawned dead zone, reached a peak size of over 8,400 square miles (21,800 square km). Over the years, upstream flooding and annual runoff along the Mississippi River, as well as storm surges from the monstrous category 4 (winds 131-155 mph (114-135 kt)) Hurricane Katrina in 2005, transported a heavy load of organic and inorganic compounds and toxic chemicals into the Gulf of Mexico. On top of this, British Petroleum's Deepwater Horizon drilling rig exploded off Louisiana on April 20, 2010. The resulting massive oil spill threatened the well-being of a myriad of marine species, including fish and shellfish, but did not result in a record-size dead zone[7]. Although it was heartening to see daily news coverage of a marine ecosystem, following the disaster, through September 19, 2010 (when the well was officially sealed), the spill of more than 4 million barrels of oil, and the toxic chemical dispersants that were used to break up and dissipate it, will continue to impact the Gulf of Mexico for a long time. Regardless of efforts to reduce nutrient loads in the Mississippi River and ultimately reduce the size of the hypoxic zone in the Gulf of Mexico, the dead zone persists, and is far from unique.

Over the last 50 years, the world has experienced a nearly exponential growth in marine dead zones, largely as a result of human-caused, nutrient enrichment, especially the overuse of inorganic fertilizers that run off agricultural soils and enter river systems. Unfortunately, dead zones are becoming more common, although generally transient, in many parts of the world. For example, due to ocean current shifts, a lack of normal upwelling, and global climate change, from 2002 to 2010, a sizable, seasonal, dead zone has developed in the Pacific Ocean, off the central Oregon coast[8] -- due west of my home. The extent of the dead zone varies from year to year. At its worst (in 2006), the hypoxic conditions severely impacted local commercial crab fishermen who pulled up their strings of crab pots expecting a healthy harvest but instead found hundreds of dead crabs.

Recent evidence has shown that increased carbon dioxide (CO_2) emissions, produced by burning fossil fuels, are likely responsible for increasing the acidity of ocean waters. The buffering capacity of the ocean may have been exceeded. The net result is that, within the ocean, calcium may not be as readily available to organisms that need it. This is particularly serious for corals that built their reefs from calcium carbonate; and for clams and conchs and allied shellfish and invertebrate species that built their shells using calcium carbonate.

A gripping and direct explanation of several of the current environmental threats and degradations facing marine life and the Earth's oceans is presented by Alanna Mitchell in *Seasick: Ocean Change and the Extinction of Life on Earth*.[9] For example, environmental stresses caused by excessive CO_2 emissions are contributing to the worldwide decline of coral reef ecosystems in at least two ways: through increased ocean temperatures (as a result of the greenhouse effect), and through increasing ocean acidity. Coral reefs are big, and hence their decline is easily visible. Yet, scientific evidence suggests that increased CO_2-emission-based ocean acidification also causes quite serious damage to very small life forms that graze algae and provide a broad nutritional base for many marine food webs.

Although ocean acidification might seem to be a problem of the future, it has already started to wreak havoc with shellfish. Oyster hatcheries on the Pacific coast of Oregon have experienced several cohort failures over the last 5 years. By carefully examining ocean chemistry, scientists were able to determine that failure occurred when larval stages were exposed to naturally occurring high CO_2 levels in recently upwelled seawater, during their first 48-hours, a critical period for shell construction.[10] By pro-actively monitoring water chemistry, and isolating larvae from dangerous levels of ocean acidity, oyster seed (spats) can be produced in tanks in hatcheries, however larvae that encounter excessively acidic seawater in the wild would almost certainly fail to survive.

In a somewhat controversial move, a cadre of international scientists, led by Boris Worm, have recently predicted that the world's seafood supply could disappear by 2048, if species continue to decline at the current rate.[11] They defined the collapse of fisheries as the point when fish catches drop below 10 percent of their historic levels. Based on this definition, they reported that stocks of 29 percent of all fished species had already collapsed by 2003. This prediction has caused a lot of defensive denials by the seafood industry, and rebuttal of their modeling techniques by other fisheries scientists. However, whether their prediction is true or not, it has opened an important dialogue about the potential collapse of seafood populations, worldwide, and created an appreciation of the fact that we need to make responsible choices when we eat seafood.

What are the benefits of including seafood in a diet?

Around the world, seafood represents a primary source of relatively inexpensive, animal protein for millions of people. Although the amount and type of fish that is consumed varies from region to region, as well as between developed and still-developing countries within a region, seafood essentially nourishes much of humanity. For centuries, fish provided a cheap protein source for industrial workers, and a subsistence diet in poor coastal villages. However, by the end of the twentieth century, rising prices and the economics of commercial fisheries had morphed the image of fish from the food of poor people into the highly sought-after, expensive, luxury food of the affluent. Wealthy clientele will gladly pay outrageous prices for seafood delicacies such as the lips of Napoleon wrasse in fancy restaurants in Hong Kong (at $250 a pair), exquisite bluefin tuna sushi in Tokyo ($75 per serving), or shark fin soup in Hong Kong (up to $100 for a cup). However, the majority of people eat fish that is lower on the food chain, and pay much less for it.

Fish and shellfish can be important components of a well-balanced, healthy diet. From a nutritional standpoint, seafood is outstanding. It is a wholesome and healthful food choice. As a result, a large number of current nutritional guidelines recommend consuming seafood at least once a week. Fish and shellfish not only provide significant sources of easily digestible, high-quality protein and other essential nutrients, they are low in saturated fat, and rich in omega-3 fatty acids. A single serving (6 ounces /150 grams) of seafood can easily provide 50 to 60 percent of an adult's daily protein requirement. A serving of fish is lower in calories and fat than meat or poultry, due to its high moisture content, and low fat content. However, not all seafood choices are nutritionally equal. For example, lean, white fish are an excellent source of extremely low-fat animal protein, while darker, oily fish contain large amounts of beneficial omega-3 fatty acids.

Seafood also represents a good dietary source of minerals and vitamins.[12] Briefly, these differ in chemical composition. Minerals are simple chemical elements, and vitamins are more complex, chemical compounds. While oxygen, carbon, hydrogen, and nitrogen form the chemical building blocks of our bodies (comprising the bulk of body weight), we also need relatively large amounts of several minerals. In terms of body requirements, these are considered "major" minerals. Our bodies require the greatest amount of calcium. Phosphorus, potassium, magnesium, sodium, chloride, and iron comprise the rest of the "major" minerals. Others essential minerals that we require in relatively small amounts are considered trace minerals.

Calcium is the most abundant mineral in our bodies. It is stored principally in our bones and teeth. It keeps bones healthy and strong, and may also prevent high blood pressure, heart attack, and colon cancer. Calcium that circulates in the blood stream facilitates hormonal and enzymatic activities that regulate energy release, metabolism, and digestion. It also plays a role in nerve communication, muscle contraction, and promotes blood clotting. The bones of canned fish are a good source of calcium, as the canning process dissolves the bones. Small, soft, chewable bones of canned fish like sardines and anchovies, can be eaten without danger. Larger bones in canned fish, such as the vertebrae of salmon and mackerel, can be easily crushed (with a fork or a spoon) to more evenly distribute their calcium throughout the fish meat.

Phosphorus is the second most abundant mineral in our bodies. Working in conjunction with calcium, at a ratio of one part phosphorus to two parts calcium, it builds strong bones, and teeth. At the cellular level, in the form of adenosine triphosphate (ATP), phosphorus plays a major role in turning the energy stored in cells into usable fuel. Phosphorus plays a role in regulating blood pH (acid-base balance). It also strengthens cell walls, and facilitates the solubility of fats, and nutrient transport. Phosphorus is abundant in seafood.

Functioning as an electrolyte (an ionic solution that facilitates cellular chemical reactions), potassium is necessary for general health and regulating blood pressure. It plays an important role in balancing body water and regulating pH, as well as in nerve function. It aids in the conversion of blood sugar into an energy form that can be stored, in the liver and muscles, and accessed when needed. Potassium may also protect against heart disease, cardiac arrhythmias, kidney stones and stroke. Mussels, scallops, and clams are good sources of potassium.

Magnesium is essential for energy production and nerve function. It activates enzymes and is involved in protein synthesis. It reduces heart disease, and stress, and is useful in treating chronic conditions such as high blood pressure, fibromyalgia and diabetes. Nuts and cereal grains contain the highest concentrations of magnesium. However, seafood contains more magnesium than equivalent-sized portions of most meat products.

Although sodium is a necessary mineral, it attracts water. Sodium plays a role in body water and acid-base balance, as well as in nerve function. High levels of sodium in our tissues result in fluid retention and increased blood volume. This in turn forces the heart to work harder, and raises blood pressure. Most of us consume more sodium than our bodies need. Sodium occurs in common salt (sodium chloride). For optimal blood pressure, the best food choices are those that are low in sodium. Fresh seafood is naturally low in sodium.

Chlorine is a vital component of the acid that breaks down food in our stomachs (hydrochloric acid). It is also an important constituent of spinal fluid. Chlorine is the "chloride" component of common salt. We seldom need to worry about getting enough chlorine, unless we are on a sodium-restricted diet, or lose excessive amounts of body fluids, through illness or by sweating. Seafood products that are processed with salt, such as canned or smoked fish, contain relatively large amounts of chlorine.

Iron is essential for red blood cell formation, and is a major component of hemoglobin, the oxygen-carrying pigment in red blood cells. Hemoglobin supplies oxygen to the body, and carries away carbon dioxide wastes. Similarly, myoglobin, an iron-containing pigment in the muscles, supplies oxygen to muscles. Iron also performs important functions in immune responses. Oysters, clams, mussels, and dark-fleshed fish such as bluefish and sardines are good sources of iron.

Copper is one of the most abundant trace elements in our bodies. It is an essential element in many enzymes and proteins. Copper maintains energy levels, facilitates the absorption of iron, is necessary in pigment production, and is important in the creation and maintenance of collagen (a protein that is the constituent of connective tissue and bone). It is also essential in antioxidant production, and has anti-inflammatory properties. Oysters, crab, and lobster are good sources of copper.

Iodine is essential for proper thyroid function. It facilitates weight loss through burning extra body fat, promotes growth, maintains energy levels, and helps builds healthy skin, hair, nails and teeth. In developed countries, most dietary iodine is obtained from iodized salt. Other important sources of iodine include shellfish and sea vegetables such as kelp and seaweed.

Selenium is a trace element that occurs in nearly every cell. It helps boost the immune system. Perhaps most notably, it acts as an antioxidant, and may discourage tumor growth. Foods that are high in selenium include tuna, oysters, flounder, and shrimp.

Zinc promotes the production of white blood cells by the thymus gland, as an essential component of healthy immune response. Zinc has anti-inflammatory properties and protects against fungal and bacterial infections. Oysters and crustaceans (i.e., shrimp, lobster, crab) are good sources of zinc.

Niacin (vitamin B3) is essential to maintain healthy skin and regulate metabolism. It is a natural, cholesterol-lowering vitamin that has been used for over 50 years to improve cholesterol levels. It promotes circulation as well as nerve function. Tuna, swordfish, halibut, and salmon are all good sources of niacin.

Vitamin D (calciferol) is a fat-soluble vitamin that maintains proper blood levels of calcium and phosphorus, and is very important in building and maintaining strong bones. Not many foods are rich in vitamin D, and as a result many are fortified with it. However, fish oils, and fatty fish, such as salmon, mackerel, and sardines, are good, natural, dietary sources of vitamin D.

Over thousands of years, Asian cultures have realized the health benefits of incorporating sea vegetables (seaweed) into their diet. In addition to being essentially fat free, low in calories, and high in fiber, seaweeds can be considered as exquisitely balanced natural foods: mineral-rich, multi-vitamin, whole foods from the sea. For as long as rain has fallen on Earth, all of the nutrients and minerals that have been washed out of rocks and soil have flowed downstream, eventually ending up in the sea. Sea vegetables have ready access to this vast supply of minerals in the sea, as a result, they contain all the major minerals that are required by humans. Their content of essential minerals surpasses land plants. Edible seaweeds also contain many of the trace elements, and many essential vitamins including vitamin A, vitamin B complex (B1, B2, B3 [niacin], B5 [pantothenic acid], B6, and B12), vitamin K, and folic acid. A number of the health benefits that are attributed to seaweed result from its relatively high iodine content. These include the promotion of healthy thyroid hormone function, regulation of metabolism, and nourishment of the skin and hair. The high mineral content of seaweed may protect against heart disease and hypertension. Additionally, seaweed is thought to provide relief from arthritis. And seaweed in the diet may help the body fight infections and eliminate toxins in the digestive system.

Seaweeds are usually dried after harvesting, and are then packaged and shipped around the world. One of the most recognizable forms of edible seaweed is nori, the thin rectangles that are wrapped around sushi rolls. Other edible seaweeds include kelp, kombu, wakame, arame, and dulse, which are used to make salads, can be added to soups, and stews, or can be cut into thin strips and sprinkled on other prepared dishes. Dried seaweed products are most readily available at Asian markets and health food stores, although many large grocery stores stock some seaweed products in their Asian or health food departments. For those who opt to harvest their own sea vegetables: learn how to identify what's out there, and remember that seaweeds can take up heavy metals, so choose a foraging site that is fairly remote from sources of industrial or urban pollution.

Although seafood is not a wonder drug, the long list of health benefits attributed to a seafood diet can make us wonder. Eating seafood may improve intelligence, skin condition, and migraine headaches. A seafood diet may relieve stress, depression, and schizophrenia. It may decrease the risk of cancer, diabetes, asthma, Alzheimer's disease, and heart disease. And it may prevent obesity. A diet rich in seafood appears to be beneficial for early childhood brain development, and may have a positive influence on dyslexic and hyperactive children. This litany of health claims sounds like something a snake oil salesman might boast, however substantial evidence backs up these claims, and many of the health benefits are specifically related to fish oil rather than snake oil.

What are the health benefits of omega-3 fish oils?

For years, dietary fats had a reputation as being bad, and many diets promoted the consumption of low-fat and fat-free items. However, it has recently become evident that there are "good" fats as well as "bad" fats. Saturated fats have been shown to contribute to heart disease by raising bad cholesterol (Low Density Lipoprotein or LDL) levels, and clogging arteries. They are considered "bad" fats that should be avoided, for the sake of cardiovascular health. Most recently, transfats have been given a "bad fat" designation. Transfats are man-made, processed, hydrogenated fats that have a long shelf life. The hydrogenating process destroys essential fatty acids and produces trans fatty acids (transfats) that are lacking in nutritional value. Health-wise, transfats are bad because they raise bad cholesterol (LDL) levels and reduce good cholesterol (High Density Lipoprotein or HDL) levels. The damage that they cause can lead to stroke, obesity, and type 2 diabetes.

Unsaturated fats are considered good fats, in terms of their role in reducing coronary disease. However not all unsaturated fats are nutritionally equal. Recent studies have shown that the amount, type, and composition of dietary fat all influence the development of cancer. For example, dietary omega-6 polyunsaturated fatty acids have been shown to facilitate and stimulate the growth of cancer tumor cells, particularly breast cancer. It is not really possible to reduce cancer risks by eliminating omega-6 fatty acids from the diet, because they are essential for some biochemical functions and the maintenance of overall health.

A very "good" fat would be one that provides protective benefits against both coronary disease and cancer. Research has confirmed that omega-3 polyunsaturated fatty acids protect the cardiovascular system via reducing coronary disease, and reduce rates of cancer, and are thus a very "good" fat.[13] There are essentially three types of omega-3 fatty acids: alpha-linolenic acid (ALA), eicosapentaenoic acid (EPA), and docosahexaenoic acid

(DHA). The simplest, ALA, is synthesized by plants, including marine algae. In the ocean, the water column contains a multitude of small plant and animal species that drift with ocean currents. Collectively, these organisms are known as plankton (derived from the Greek term for wandering). Within the plankton, small invertebrate animals called zooplanton eat marine algae called phytoplankton (zoo- and phyto- denote animal and plant, respectively) and synthesize the longer EPA and DHA molecules from ALA. The two longer, animal-made molecules (EPA and DHA) appear to provide the greatest health benefits. Fish that ingest zooplankton concentrate these omega-3 fatty acids at each trophic level. The flesh of predatory fish that ingest fish, that have ingested zooplankton, contains higher concentrations of omega-3 fatty acids than their prey did.

In humans, the consumption of fish oils that are rich in long-chain, omega-3 fatty acids is extremely beneficial in the regulation of blood clotting and blood vessel constriction. For older people, omega-3 fatty acids appear to reduce sudden cardiac death through beneficial effects on the electrical functioning of the heart: the heart's rhythm and pace. Consumption of omega-3 fatty acids may also reduce risks of metabolic disorders, reduce incidence of asthma, reduce the development of psoriasis, reduce the pain and inflammation of arthritis, and enhance immune response. Omega-3 fatty acids may also play a beneficial role in treating cardiac arrhythmias, depression, and irritable bowel syndrome. Heart-healthy, omega-3 fish oil supplements are not a miracle drug. People with advanced heart disease cannot be cured simply by taking them. However, many people will experience a variety of health benefits from including oily fish, or omega-3 fish oil supplements, in their diet.

In Europe, patients are routinely prescribed purified, long-chain, omega-3 fish oils after heart attacks. However, in the US the Food and Drug Administration (FDA) has not approved prescription-grade fish oil for treatment of heart conditions, but has licensed it only for the treatment of high blood triglyceride levels. Instead in America, fish oils are relegated to the aisles of nutritional supplements, which can be a twilight zone of quality control, and cardiologists generally prescribe more expensive pills or invasive procedures for their patients. The under-appreciation of fish-oil supplements in the US is evident from the fact that although, in 2006, the American Heart Association recommends the consumption of two servings a week of fish rich in long-chain, polyunsaturated, omega-3 fatty acids, to promote cardiovascular health, it recommends that patients with coronary heart disease should only take fish oil supplements (at a dosage of one gram per day) under the supervision of a physician.[14] Because the FDA hasn't approved prescription-grade fish oil for the treatment of heart conditions, any American physician who follows these guidelines and prescribes the purified, long-chain, omega-3 fish oils that are routinely used in Europe, is doing so outside the FDA-recommended uses of the product.

Although the claim that "fish is brain food" sounds a little extreme, brain cells require good fats like the long-chain omega-3 fatty acids, EPA and DHA, that are found in fish. Optimal brain development in infants and small children has recently been correlated with maternal consumption of high amounts of seafood during pregnancy. Beyond childhood, our brains still benefit when we eat fish. Brain cell membranes and the tissues enclosing nerves are composed principally of fats and water. Eating fish quite likely facilitates brain chemistry. Omega-3 fatty acids also likely improve brain function through their anti-inflammatory effects and through their benefits to cardiac function. A diet that is good for the heart provides a better supply of blood and oxygen to the brain. Recent evidence also suggests that eating fish may slow the decline of memory and cognitive skills that frequently occurs in old age.

Omega-3 fatty acids are important in visual development in infants. In adults, a diet rich in omega-3 fatty acids provides protective benefits against macular degeneration and dry eye syndrome. Nutritional supplements that contain omega-3 fatty acids are used as one treatment for dry eye syndrome.

One easy way to eat foods that are rich in omega-3 fatty acids is to regularly consume fatty fish such as herring, mackerel, salmon, trout, and tuna. The benefits of this type of diet have been observed in long-term studies by Danish researchers, H.O. Bang and Jorn Dyerberg (published in the 1970s), that tracked disease in populations of Greenland Eskimos. These studies examined the factors responsible for the presence or absence of heart disease. The very low incidence of cardiovascular disease in their population was correlated with their predominantly seafood diet. Greenland Eskimos do not possess some special genetic protection against heart disease. This has become clear when Greenland Eskimos have adopted western diets. The unfortunate result of changing their diet is that their incidence of cardiovascular disease has increased.

Omega-3 fatty acids may also be obtained through purified fish oils, such as cod liver oil. However, not all processed fish oils are equal. Two basic problems arise in the processing of fish oils: removing contaminants and stabilizing the oil to prevent oxidation which results in unpleasant tastes and odors. Impurities in fish tissues/fish oils include two types of man-made chemicals: organochlorine pesticides, such as dichlorodiphenyltrichloroethane (DDT) and hexachlorohexane (HCH); and polychlorinated biphenyls (PCBs). Resisting both biotic and abiotic degradation, DDT and PCBs are recalcitrant compounds that persist in the environment long after their use has been discontinued. These chemical compounds are soluble in lipids, and as a result, fatty fish have higher levels of these compounds than leaner fish.

Heavy metals are another type of impurity found in fish tissues. Concentrations of heavy metals such as mercury, lead, and cadmium in aquatic environments have increased dramatically in the last few decades. These metals occur in various forms in the marine environment. Free cadmium ions are taken up by fish. Methylated molecules (i.e., molecules with a methyl radical, CH_3, attached) of mercury and lead are synthesized by algae, and these forms are taken up by fish. The lipid soluble nature of these compounds (methyl mercury and tetramethyl lead) results in their concentration in lipid-rich, or fatty, tissues. Metal content of fish oil is influenced by the level of contamination of the fish (the raw material being processed) as well as the amount of time the fish tissues are stored prior to processing. The processing of marine fish oils for human use removes much of the man-made and naturally occurring contaminants, and as a result, processed fish oils contain lower levels of these chemicals than would be expected based on their concentration in fish tissues.

One caveat should be added. Although it is tempting to consider taking a daily omega-3 fish oil supplement rather than obtaining long-chain, omega-3 fatty acids by eating fish, consider how many fish were processed to yield a single bottle of supplements. In most cases, hundreds or thousands of small oily fish, such as anchovies or menhaden, were run through rendering plants to yield one bottle of purified omega-3 fish oil supplements. Unfortunately, despite sacrificing all these little fish, after taking one capsule, we still need to eat dinner. When we obtain omega-3 fatty acids through diet, we fill our stomachs at the same time. Both ecologically and energetically, eating fish makes more sense than swallowing fish oil capsules. The maximum nutritional value of fish is realized when humans eat fish directly. Theoretically, we could use most of the fish in the sea to process for fish oil supplements, but the world would be poorer and hungrier as a result.

How are heavy metals biomagnified through the food chain?

In today's "global" economy, pollution, an unfortunate by-product of globalization, has spread beyond the limits of civilization. Just as massive Asian dust storms carry industrial pollution across the Pacific and deposit it on North American soil, ocean currents have transported waste products and pollutants great distances to remote sites such as Antarctica. Additionally, spawning salmon[15] and migratory seabirds[16] have recently been shown to transport marine-derived contaminants into the arctic. Thus, former bastions of pollution-free water now are tainted with traces of industry and agriculture. Although this creep of globalization may not presently be at toxic levels, any increases should be viewed with alarm rather than acceptance.

One of the biggest questions that we face is which seafoods are safe to eat? Many of us are particularly concerned about tuna, in light of recent reports of high mercury levels discovered in canned tuna, and fresh tuna from fish markets. The take home message that most people retain, is avoid tuna, because it is high in mercury content. However, not all tuna is high in mercury, and some other fishes may also be high in heavy metal concentrations.

Mercury is not the only heavy metal that is present in fishes, but its level is frequently measured due to the significant health hazards that it represents to infants, young children, and women who are pregnant or who may later bear children. Infants who are exposed to high doses of mercury suffer principally neurological effects, as well as developmental delays and depressed intelligence.

Mercury and other heavy metals accumulate via the food chain. In the simplest of terms, energy coming from the sun is captured by plants, which then convert solar energy into a form that is available to animals, via photosynthesis. Small, plant-eating animals are eaten by larger animals, which are in turn eaten by larger animals. Each step in this food chain is called a trophic level. Terrestrial food chains can be fairly short. For example, deer consume plants, and are preyed upon by cougars. This represents 3 trophic levels: plants are the primary producers, deer are herbivores, and cougars are carnivores. However, marine food chains often consist of more than 4 or 5 trophic levels. In many marine food chains, single-celled algae (plants) are the primary producers (level 1), and small zooplankton species that eat the algae function as herbivores (level 2). They are often eaten by larger zooplankton species (level 3), which in turn are ingested by small fish (level 4). The small fish are then consumed by larger fish (level 5), which might be eaten by a large predator (level 6). In an aquatic environment, bacteria and algae transform inorganic mercury into methyl mercury which is extremely toxic, and readily absorbed, when it is ingested.

In the ocean, phytoplankton (i.e., algae) and bacteria take up small quantities of the heavy metals. These metals are not readily metabolized, that is, they aren't easily burned up through respiration. At the next step in the food chain, small zooplankton species (such as rotifers or copepods) ingest the bacteria or algae, and the small amounts of heavy metals contained in their tissues. Because the zooplankton also cannot metabolize the heavy metals, they are concentrated in the animal's tissues. Small zooplankters are eaten either by larger zooplankters (such as shrimp) or by small fish, which are then eaten by larger fish, continuing up the food chain. At each trophic level in the food chain, the bulk of ingested heavy metals are neither burned up, nor are they readily eliminated as waste products. Instead they are concentrated in the predator's tissues. This process is called biomagnification or bioaccumulation. The levels of heavy metals (measured in parts per million, PPM) increase, or are magnified, at each trophic level.

As a result of biomagnification, tissues of the top predators in the food chain, like tuna, swordfish, and shark often contain very high levels of heavy metal contamination. Tissues of a very old, very large, top predator would likely contain the highest levels of heavy metals.

In the case of canned tuna, large-scale commercial canners prefer processing very large tuna, to minimize processing costs. These big, old tuna are top predators, and contain high concentrations of heavy metals. Smaller, "boutique" canneries generally prefer smaller, younger tunas. As a result, their canned tuna fish contains lower levels of heavy metals.

Some confusion exists relating to the fact that tuna is a collective term, and not just one species. Rather, several species of tuna are commercially harvested. The biggest, and perhaps best known, is bluefin tuna, which can reach hundreds of pounds/kilograms. Yellowfin tuna, little tunny, skipjack tuna, and albacore are also harvested.

The best possible situation for assessing heavy metal biomagnification risks is to know the fish we eat: know its size as well as where it was caught. In a perfect world, we might catch it ourselves. However, faced with the realities of life in the twenty-first century, most of us don't have the time or resources to catch all the fish that we consume. One option is to know the fisherman who caught it. Go to the coast and buy fish off the fishing boats. Local extension offices may be able to provide guidance as to when and where various fish are harvested. Along the Oregon coast, during the summer, commercial fishermen routinely sell albacore directly to consumers. Many consumers buy large quantities to can, smoke or freeze. This has the added benefits of putting more money directly into the pockets of local fishers, and saving the buyer money. Some of the fishermen market their own "boutique" value-added products after canning or smoking their catch. Often these products are more expensive than comparable mass-marketed products, but they are healthier both for consumers and the environment, because they contain less mercury and because their capture is usually done via hook and line, which maximizes catch quality and minimizes the incidental capture of something other than the principal target of the fishery.

If you don't live close to the sea, ask questions of your fish monger or your server in a restaurant. Learn as much as possible about what you eat. Take knowledge with you as you purchase fish, and don't be afraid to ask questions.

In addition to tuna, other tasty, long-lived, top predators also may be loaded with heavy metals: including swordfish, marlin, sturgeon, shark, king mackerel, and tilefish. If we choose to eat them, we shouldn't make a steady diet of them. Eat them infrequently. Contaminants, especially heavy metals such as mercury, can be particularly damaging when consumed by infants, small children, pregnant women, breast-feeding women, women of childbearing age that are likely to become pregnant, and people suffering from liver or kidney damage. All these groups are considered at risk and should restrict their intake of seafood that is likely to be laden with contaminants. Guidelines and fish consumption advisories have been established by the Environmental Protection Agency (EPA), as well as state and tribal authorities for bodies of water under their jurisdiction. For sport-caught fish, local health departments may also provide information about local health advisories.

In March 2004, the EPA and Food and Drug Administration (FDA) jointly issued three recommendations for the consumption of fish and shellfish by a sensitive segment of the population who are at great risk from mercury in fish. This advisory was specifically aimed at pregnant women, women who might become pregnant, nursing mothers, and young children:[17]

1. Do not eat shark, swordfish, king mackerel, or tilefish due to their high levels of mercury.

2. Eat up to 12 ounces (340 gram) of seafood a week (2 meals), choosing varieties that are low in mercury: shrimp, canned light tuna, salmon, pollock, and catfish.

3. Check local health advisories for recreationally caught fish. If advisories are not available, limit consumption to 6 ounces (170 grams), and avoid eating other fish during the week.

One take-home message here is that when we are considering eating a top predator, for reasons of biomagnification of heavy-metal pollutants, such as mercury, size absolutely matters. In the big fish versus little fish decision, smaller is better. Choose to eat the smaller fish. While age may bring wisdom, with long-lived fish, age brings a higher concentration of contaminants and toxins.

In August 2004, the EPA released a fact sheet, a national listing of fish advisories, regarding the consumption of fish caught in freshwater rivers and lakes.[18] Nearly one quarter of rivers and one-third of lakes in the US contain sufficient pollutants that people should limit their consumption of fish caught in these waters. These warnings covered about 35 percent of lake acreage and 24 percent of river mileage in the US. Almost all these advisories were related to pollutants such as mercury, dioxins, polychlorinated biphenyls (PCBs), pesticides, and heavy metals such as arsenic, lead, and copper. Coal-burning power plants are one of the primary sources of mercury pollution in these river and lake systems. This is an example of how air pollution results in water pollution which in turn poisons the food that we eat. The interconnected contamination of air, water, and food is particularly scary in light of efforts by anti-environmental politicians in the federal government to roll back clean air and water standards, to benefit energy companies and industry.

Air pollution from other continents in our "global village" can also be transported across oceans and then rain contaminants down upon distant watersheds. Scientific evidence suggests that long-range atmospheric transport of Asian air masses contributes substantially to the deposition of mercury in western rivers in the US. Thus the concentration of mercury in fish tissues across the American West is likely related in part to Chinese air pollution.

For a number of reasons, including protecting public health and following the dominant pathway by which humans are exposed to methylmercury, the EPA has determined that measuring methylmercury concentration in fish tissues was a more useful water quality criterion than measuring mercury concentration in water, directly. In a January 2001 report, the US EPA set 0.3 micrograms of mercury per gram of fish fillet as the acceptable limit for human consumption that should not be exceeded.[19]

A recent survey of over 600 western rivers and streams examined mercury concentration in fish tissues.[20] The survey found widespread mercury contamination of fish. Not surprisingly, large fish (≥ 130 mm {5 inches}) had higher levels of mercury in their tissues than small fish, and piscivorous fish (i.e., fish-eating fish) had higher levels than non-piscivorous fish. Salmonid fishes (members of the family Salmonidae, which includes salmon and trout) were the most common fish in the survey. They were also observed to have the lowest body burdens of mercury of all game fish in the survey. The salmonids in this survey, brook trout (*Salvelinus fontinalis*), brown trout (*Salmo trutta*), cutthroat trout (*Oncorhynchus clarki*), and rainbow trout (*Oncorhynchus mykiss*), were all non-piscivorous, and their tissues exceeded the concentration of mercury that EPA established as its water quality criterion for protecting human health from methylmercury along only 2.3 percent of the assessed stream length in which they were observed. Other game fish such as northern pike (in the genus *Esox*), bass (in the genus *Micropterus*), and walleye (in the genus *Sander*) were less abundant than salmonids in this study, but their tissues exceeded the EPA mercury threshold along 57 percent of their assessed stream length. The take home point here is that these wild trout would be considered very safe to eat, and the other game fish (northern pike, bass, and walleye) would not be a wise seafood choice.

Other persistent contaminants

Polychlorinated biphenyls (PCBs) represent a category of over 200 man-made chemical compounds that share structural and physical properties. They are essentially colorless, odorless compounds that are composed of 2 benzene rings, with 2 or more of the hydrogen atoms replaced by chlorine atoms. The flame-resistant property of PCBs, and the fact that they don't break down and are chemically inert, made them an appealing choice as electrical insulators used in transformers, capacitors, and other electrical equipment. They were also used as fire retardants in a variety of products ranging from bread wrappers and cereal boxes to heating coils, industrial drill lubricants, and the caulking compounds used in skyscraper construction. From 1930 to 1970, 1.4 billion pounds (0.6 billion kg) of PCBs were produced in the United States.

Although the manufacture of PCBs was banned in the United States in 1989, the replacement of PCB-containing products was not mandated at that time. As a result, some PCB-containing products may still be in use. Although, the vast majority of PCBs in the environment are the result of past releases of the compounds, due to their recalcitrant nature, high quantities of PCBs are still present in our environment. They have entered the atmosphere through smokestacks and burning of products that contained PCBs. They have entered the water supply through sewers, leakage from PCB-containing products, and land-fill leaching. They have been detected in air, soil, surface water, sediments, plants and animals in all regions of the planet. PCBs are not very soluble in water, nor very volatile. As a result PCBs accumulate in sediments. The PCB-laden sediments then act as an environmental reservoir from which PCBs can be released over a long time span via physical disturbances of the sediments, such as dredging. The environmental half-life of PCBs may be up to 5 years.

Unfortunately, some of the same properties that made PCBs so attractive in chemical terms, make them very toxic, and they accumulate in animal tissues. While not soluble in water, PCBs are soluble in fat. This fat solubility results in their accumulation in animal tissues via aquatic food chains. Biomagnification or bioaccumulation of PCBs occurs through the food chain, in the same way that mercury is concentrated from one trophic level to the next. PCBs have been detected in both freshwater and marine fish and shellfish, with the amount depending on a number of factors including, size, feeding ecology, exposure, life history, feeding grounds, and fat content. PCBs occur in food chains far from where the compounds were manufactured or used. The detection of PCBs in blubber tissue of sperm whales that had stranded on beaches in Europe is an example of the alarming reach of these contaminants. Although these whales live far from continents, and feed in the deep ocean, on fish and squid, man's activities have tainted them. Tissues of another top predator, this time in the Arctic food chain, the polar bear, have also been shown to contain PCBs.

Acute exposure to high levels of PCBs (from industrial accidents) has resulted in eye irritation, headache, fatigue, skin lesions, and digestive disorders including nausea and vomiting, as well as liver dysfunction. Chronic exposure to low levels of PCBs may result in developmental problems in children, liver damage and various cancers.

To minimize the risk of PCB exposure, the first step is to pay attention to local fishing advisories. We should follow the same guidelines that we would for other contaminants. Don't eat fish caught near contaminated sediments. Beyond that, if the source of sport-caught fish is unknown or questionable, it is wise to exercise caution and limit the amount consumed, and eat it infrequently. Eat small, lean, and short-lived fish species, rather than large, fat, long-lived species. Skin the fish, and discard the internal organs. Don't eat raw fish. Discard fatty tissues, especially belly fat, the dark fat above the backbone, and dark fat along the lateral line. Grill or broil the fish so that the fat drains away from the fish, and discard the fat.

Dioxins and furans comprise another group of polychlorinated hydrocarbons with properties that are similar to PCBs, both chemically and physically. They persist in the environment and are biomagnified through the food chain. Although dioxins (polychlorinated dibenzo-para-dioxins) and furans (polychlorinated dibenzo-furans) are not produced intentionally, they are the byproducts of various industrial processes such as the manufacture of disinfectants and herbicides, chlorine bleaching of paper pulp, and incomplete combustion of fossil fuels, including industrial and municipal waste incineration. Small amounts of dioxins are also produced naturally by volcanoes and forest fires. Seventeen dioxins and furans have been extensively studied, and are toxic. Dioxins are highly carcinogenic to animals and are considered likely human carcinogens as well. To minimize exposure to dioxins and furans, follow the same steps as above. Avoid fish that are caught near contaminated sediments. Eat small, lean fish. Discard the skin, internal organs, and fatty tissues. Cook fish in a manner that the fish's fat drains away from the fish, then discard the fat.

When purchasing fish, the public is generally protected from contaminated fish via standards set by the FDA, and the EPA. Unfortunately, these two agencies have set different standards for recognizing PCB contamination.

To add more confusion to the issue of seafood safety, on August 24, 2004, when the EPA advisories were announced, Mike Leavitt, EPA Administrator, said "It's about trout, not tuna. It's about what you catch on the shore, not what you buy on the shelf."[21] Leavitt went on: "This is about the health of pregnant mothers and small children; that's the primary focus of our concern." The "it's about trout, not tuna" announcement is particularly perplexing to consumers who have paid close attention to the EPA/FDA guidelines for pregnant women, women who may become pregnant, and small children to limit their consumption of albacore (white tuna) to 6 ounces (170 g) per week and canned light tuna to 12 ounces (340 g) per week.[22] Despite widespread knowledge that large tuna (especially bluefin tuna) contain high levels of mercury, the immensely powerful American tuna industry has been able to forestall the listing of any tuna species as fish to avoid due to mercury content. How can we not be confused?

Do the benefits of eating seafood during pregnancy exceed the risks?

While the American Food and Drug Administration (FDA) and the Environmental Protection Agency (EPA) have jointly advised pregnant and breast-feeding women to not eat shark, swordfish, king mackerel, or tilefish due to their high levels of mercury, and to limit their consumption of seafood varieties that are low in mercury to 12 ounces (340 grams) per week, to protect the brain development of their unborn children from environmental pollutants, in Great Britain, the Food Standards Agency merely advises pregnant and breast-feeding women to avoid shark, swordfish, and marlin, and limit the amount of tuna that they eat. Many American women are so confused and anxious about the seafood advisories that they seriously consider not eating any fish while pregnant in order to minimize health risks to their unborn children. However, this is not really a good strategy, because new evidence shows that during pregnancy, the benefits of eating most types of fish and seafood outweigh the risks.

In February 2007, results from the Avon Longitudinal Study of Parents and Children (ALSPAC, also known as "Children of the 90s") were released.[23] The aim of this study was to identify factors that could improve childhood health and development. This study followed nearly 9,000 women and their children for eight years, in Great Britain. One important aspect of this study was to examine the effect of seafood consumption during pregnancy on neurological development through childhood. The headline of an ALSPAC press release summarized a very important finding: Research shows that the benefits of eating seafood during pregnancy outweigh the risks. Researchers examined the amount of fish consumed by pregnant women in relation to the behavior and development of their children over eight years. Women who ate more than 12 ounces (340 grams) of fish per week while pregnant (i.e., women who exceeded the amount recommended by American guidelines) gave birth to children who 1) as toddlers, were developmentally more advanced, in terms of motor skills, communication, and social skills; 2) at age 7, demonstrated more positive social behavior; and 3) at age 8, were less likely to have low verbal IQ scores. Women who ate no seafood while pregnant gave birth to children who were significantly more likely to demonstrate deficits in motor skills, communication, social skills, and verbal IQ. At 18 months, their children were 28 percent more likely to have poor communication skills. At 42 months (3.5 years), their children were 35 percent more likely to have poor motor coordination. At age 7, their children were 44 percent more likely to have poor social behavior. And at age 8, their children were 48 percent more likely to have a relatively low IQ score, when compared with children of women who ate more than the amount of seafood recommended by US guidelines, while pregnant (i.e., ate more than 12 ounces (340 grams) of fish per week). In their developmental assessment, children of women who ate some seafood during pregnancy, but less than 340 grams per week scored between the "high seafood" and "no seafood" groups. For some comparisons, the developmental scores for the "moderate seafood" group were significantly better than the "no seafood" group.

Results from the British ALSPAC study clearly show that maternal consumption of more than 12 ounces (340 grams) of seafood per week during pregnancy had no adverse effect on a child's behavior or neurological development. Instead, consuming more seafood than the amount recommended by US guidelines, while pregnant, appeared to have beneficial effects on a child's development. These findings were opposite to the results anticipated based on US guidelines. Although the US guidelines on how much seafood pregnant women or women of child-bearing age, that were likely to become pregnant, could safely eat were undoubtedly established to protect the neurological development of their unborn children (by minimizing exposure to methyl mercury, a neurotoxin), strictly following these guidelines appears to have the opposite effect and could be detrimental to the neurological development of children. Whatever protection from trace contaminants the EPA and FDA felt that they were affording unborn children with these guidelines was negated by the detrimental effects that children suffered without access to building blocks of the brain: long-chain omega-3 fatty acids from fish. In the absence of compelling scientific evidence of harm from eating seafood, pregnant women should eat more seafood, choosing varieties that are known to be low in contaminants.

While much of the heavy metal contamination in fish comes from man-made causes, it should be pointed out that many naturally occurring seeps of heavy metals also occur. It may seem tempting to conclude that coastal waters are rather heavily impacted by agriculture and industry, and then seek the purity of high mountain lakes as a clean alternative source of fish. However, Oregon's East Lake is one example of why this isn't always a good idea. East Lake is located inside the caldera of Newberry Volcano (Newberry National Volcanic Monument) at an elevation of 6,381 feet (1945 m). Snow melt and natural springs supply water to the lake. This sounds idyllic and pure. However, since the mid-1990s, the state of Oregon has issued an advisory against eating brown trout 16 inches (405 mm) or longer caught in East Lake, and limiting the consumption of brook trout smaller than 16 inches and all other fishes caught in East Lake, due to high mercury levels from naturally occurring sources, such as soils, rocks, and a hot spring, adjacent to the lake.[24]

Each state offers its own health advisories related to freshwater fish. Be informed. Armed with the knowledge of where a fish came from, a savvy consumer can make intelligent decisions regarding the health "risks" associated with eating freshwater fish.

It wouldn't hurt to follow a bit of advice from the late Julia Child: "Take food seriously to know exactly what you are eating and inform yourself, but don't go off on scare tactics."

What should we know about naturally occurring toxins in seafood?

"All natural" food has an appealing ring to it, but an "all natural" designation doesn't translate into specific, useful knowledge about the healthfulness or safety of a particular food. For instance, mercury, cadmium, and arsenic are all naturally occurring elements, but they do not represent healthful ingredients in our food. A wide variety of toxins are synthesized by plants and animals every day, in a vast array of ecosystems around the world. These toxins can be considered as natural solutions that species have evolved to overcome obstacles to their survival. Many are used as anti-predator defenses. These bioactive compounds are known as biotoxins. Despite the fact that biochemists are analyzing the pharmaceutical potential of many marine biotoxins, in their quest to develop new therapeutic and disease-fighting, miracle drugs, some marine biotoxins are just plain toxic to humans.

Marine biotoxins can produce intoxications, or poisonings, that are separate and quite different from food-borne illnesses such as botulism and scombrid poisoning that result from spoilage and improper handling or processing of seafood. These toxins are not destroyed by freezing, cooking, drying, salting, or smoking. And, unfortunately, the toxicity of seafood cannot be determined by its smell or appearance. When ingested, marine biotoxins are capable of chemically interrupting vital physiological functions in our digestive, neurological, and cardiovascular systems. Eating marine biotoxins can cause serious human illness as well as death, which is why it is important to understand how to avoid them.

Within the seas, thousands of species of microscopic algae (one-celled plants) and their ecological equivalents (collectively referred to as phytoplankton – wandering photosynthesizers) provide an extremely important function as the base of marine food chains. In their capacity as primary producers, phytoplankters convert solar energy into carbohydrate energy, through photosynthesis. However, a few of these planktonic species produce potent biotoxins. Around 30 species of dinoflagellates (photosynthesizing, single-celled, planktonic organisms that are essentially ecologically indistinguishable from microalgae) produce toxins that can make humans ill when they ingest shellfish and fish that have fed on blooms of the toxic dinoflagellates.[25] These algal blooms are frequently referred to as red tides. However, "red tide" is really a misnomer, because the blooms aren't always red, and they are not associated with tides. Massive algal blooms discolor the seawater that contains them. The discoloration is produced by the dominant pigments in the algae. Blooms can range in color from pink and violet through yellow, orange, blue, green, brown and red, depending on the algal species. Although commonly used, the term "red tide" is also misleading because non-toxic algae can bloom and discolor seawater without producing adverse effects. And blooms of toxic algae can produce adverse effects at low cell densities that do not discolor the water. Clearly, "harmful algal bloom" is a more useful and correct label than "red tide."

The term "shellfish" generally refers to aquatic invertebrate animals with shells, although some "shellfish," such as octopus, have no shells. The bulk of commercially important shellfish come from one of two major groups of invertebrates: Crustacea which includes crabs, lobster, and shrimp; and Mollusca which includes clams, conchs, octopus, and squid. Bivalve molluscs (two-shelled shellfish in the phylum Mollusca), such as clams, mussels, oysters, and scallops, are filter feeders. They pump water through their siphons; feeding on algae, bacteria, and virtually everything that they can filter out of the water. During a bloom of harmful algae, bivalve shellfish concentrate the toxins in their digestive organs. Through the food chain, whatever eats the toxic shellfish ingests a large dose of the toxin. Shellfish toxins have been implicated in causing fish kills. But here, we will focus on their effects on humans.

The consumption of bivalve molluscs such as clams, mussels, and oysters is the most common pathway for marine biotoxins to affect humans. This is because the internal organs and intestinal tracts of intoxicated animals store these toxins. Once they are extracted from their shells, bivalves are often eaten whole, whereas, crab and fish guts are usually removed and discarded, rather than consumed by humans.

At least three separate types of shellfish poisonings and one type of fish poisoning have been attributed to dinoflagellates. These poisonings are paralytic shellfish poisoning (PSP), diarrhetic shellfish poisoning (DSP), neurotoxic shellfish poisoning (NSP), and ciguatera fish poisoning (CFP). Dinoflagellates produce bio-active compounds that can be harmful to humans in a variety of ways. Several of the toxins interfere with sodium-channel functions, and basically block the transmission of nerve impulses. One interferes with the function of calcium ion channels. Of the marine intoxications, paralytic shellfish poisoning is the best understood.

Diatoms (another group of photosynthesizing, single-celled, planktonic organisms that are ecologically equivalent to microalgae) are responsible for a fourth type of shellfish poisoning: amnesic shellfish poisoning (ASP). This intoxication results from the production of domoic acid by diatoms in the genus *Pseudonitzschia*.

Three marine intoxications are potentially fatal: paralytic shellfish poisoning, amnesic shellfish poisoning, and fugu fish poisoning. The other three marine intoxications, diarrhetic shellfish poisoning, neurotoxic shellfish poisoning, and ciguatera fish poisoning, while unpleasant experiences, are rarely fatal.

Shellfish poisoning

Shellfish poisoning is not a new occurrence, although the frequency with which outbreaks occur appears to have increased in recent decades. Paralytic shellfish poisoning has been around since biblical times. A red tide appears to have been described in Exodus (7:20-21), as the first of the ten plagues of Egypt: "... and all the waters that were in the river were turned to blood. And the fish that was in the river died; and the water stank."[26] It has been suggested that the Red Sea may have been named for frequent blooms of red algae or dinoflagellates.

Jewish dietary laws forbid the consumption of shellfish based on two biblical passages: Deuteronomy 14:9-10 "These ye may eat of all that are in the waters: whatsoever hath fins and scales may ye eat; and whatsoever hath not fins and scales ye shall not eat; it is unclean unto you."[27] And Leviticus 11:9-12: "These may ye eat of all that are in the waters: whatsoever hath fins and scales in the waters, in the seas, and in the rivers, them may ye eat. And all that have not fins and scales in the seas, and in the rivers, of all that swarm in the waters, and of all the living creatures that are in the waters, they are a detestable thing unto you, and they shall be a detestable thing unto you; ye shall not eat of their flesh, and their carcasses ye shall have in detestation. Whatsoever hath no fins nor scales in the waters, that is a detestable thing unto you."[28] These Biblical proclamations may have been based on early observations of PSP. In an interesting parallel, Islamic dietary laws, which were established in essentially the same geographic area as Jewish laws, also proscribe shellfish, considering it harmful or unlawful.

Native Americans of the Pacific Northwest have known about paralytic shellfish poisoning for centuries. However, their accounts did not contain the same level of symptomatological details as an account of PSP that was included in the journals recorded during the journeys of Captain George Vancouver in the Pacific Northwest aboard the *Discovery* and her escort, the *Chatham*. The account describes an exploration mission taken by two boats full of Vancouver's men into some little bays in present-day British Columbia.

From the journal of George Vancouver[29] on June 15, 1793:

> "They stopped to breakfast, where finding some muscles [sic], a few of the people ate of them roasted; as had been their usual practice when any of these fish were met with; about nine o'clock they proceeded in very rainy unpleasant weather down the south-westerly channel, and about one landed for the purpose of dining. Mr. Johnstone was now informed by Mr. Barrie, that soon after they had quitted the cove, where they had breakfasted, several of his crew who

had eaten of the muscles were seized with a numbness about their faces and extremities; their whole bodies were very shortly affected in the same manner, attended with sickness and giddiness. Mr. Barrie had, when in England, experienced a similar disaster, from the same cause, and was himself indisposed on the present occasion. Recollecting that he had received great relief by violent perspiration, he took an oar, and earnestly advised those who were unwell, viz. John Carter, John M'Alpin, and John Thomas, to use their utmost exertions in pulling, in order to throw themselves into a profuse perspiration; this Mr. Barrie effected in himself, and found considerable relief; but the instant the boat landed, and their exertions at the oar ceased, the three seamen were obliged to be carried on shore. One man only in the Chatham's boat was indisposed in a similar way. Mr. Johnstone entertained no doubt of the cause from which this evil had arisen, and having no medical assistance within his reach, ordered warm water to be immediately got ready, in the hope, that by copiously drinking, the offending matter might have been removed. Carter attracted nearly the whole of their attention, in devising every means to afford him relief, by rubbing his temples and body, and applying warm cloths to his stomach; but all their efforts at length proved ineffectual, and being unable to swallow the warm water, the poor fellow expired about half an hour after he was landed. His death was so tranquil, that it was some little time before they could be perfectly certain of his dissolution. There was no doubt that this was occasioned by a poison contained in the muscles he had eaten about eight o'clock in the morning; at nine he first found himself unwell, and died at half past one; he pulled his oar until the boat landed, but when he arose to go on shore he fell down, and never more got up, but by the assistance of his companions. From his first being taken his pulse were regular, though it gradually grew fainter and weaker until he expired, when his lips turned black, and his hands, face and neck were much swelled. Such was the foolish obstinacy of the others who were affected, that it was not until the poor unfortunate fellow resigned his life, that they could be prevailed upon to drink the hot water; his fate however induced them to follow the advice of their officers, and the desired effect being produced, they all obtained great relief; and though they were not immediately restored to their former state of health, yet, in all probability, it preserved their lives. From Mr. Barrie's account it appeared, that the evil had arisen, not from the number of muscles eaten, but from the deleterious quality of some particular ones; and these he conceived were those gathered on the sand, and not those taken from the rocks. Mr. Barrie had eaten as many as any of the party, and was the least affected by them. This very unexpected and unfortunate circumstance detained the boats about three hours; when, having taken the corpse on board, and refreshed the three men, who still remained incapable of assisting themselves with some warm tea, and having covered them up warm in the boat, they continued their route, in very rainy, unpleasant weather, down the south-west channel, until they stopped in a bay for the night, where they buried the dead body. To this bay I gave the name of Carter's Bay, after this poor unfortunate fellow; it is situated in latitude 52°48', longitude 231°42': and to distinguish the fatal spot where the muscles were eaten, I have called it Poison Cove, and the branch leading to it Muscle Canal."

Although PSP has been known for a long time, and many coastal states and provinces actively monitor biotoxin levels in shellfish populations, unfortunately, shellfish poisoning cannot be relegated to history books. Outbreaks occur worldwide, through all seasons of the year. A 20-year survey of paralytic shellfish poisoning in Alaska (from 1973 to 1992) identified 54 outbreaks involving 117 people.[30] In June of 1990, PSP sickened fishermen in both Massachusetts and Alaska.[31] During a fishing trip on Georges Bank (east of Cape Cod, MA), six fishermen nearly died from PSP after eating a meal of steamed mussels. All were hospitalized, and recovered fully. However, in Alaska, of seven fishermen that shared a meal of butter clams that were collected on a beach on the Alaska Peninsula, six suffered PSP symptoms, and one died.

During May and June of 2011, 21 cases of PSP were identified in southeast Alaska.[32] This represents a substantial increase over the typical number of cases reported in a year.

Paralytic shellfish poisoning (PSP) is life threatening, with a rapid onset of symptoms. It causes the most severe symptoms of all types of shellfish poisonings. The first symptoms can occur within minutes: a numbness and tingling around the mouth and lips; followed by tingling around the face, and neck, and a prickling sensation in the fingers and toes. Muscular weakness follows. Nausea and vomiting can occur, but clearly neurological symptoms are the most serious. Symptoms can include difficulty swallowing, a sensation of the throat swelling, incoherence or loss of speech, complete paralysis, and respiratory failure. The primary toxin responsible for PSP is saxitoxin, produced by dinoflagellates in the genus *Alexandrium*.

The best way to prevent PSP is through large-scale programs to monitor toxins in populations of shellfish. By frequently measuring toxin levels in mussels, clams, scallops, and oysters, areas can be rapidly closed when dangerous toxin levels are reached. Commercially harvested shellfish are frequently tested, to insure their safety. Further, knowingly selling shellfish from closed areas is illegal. Tourists and people who are unfamiliar with an area make up a disproportionate share of PSP cases, either because they disregard quarantines, or are unfamiliar with local traditions ensuring the safe consumption of shellfish.

Diarrhetic shellfish poisoning (DSP) is named for the predominant symptom experienced by ill victims: diarrhea. It is a relatively mild form of shellfish poisoning. It is different from PSP in that its symptoms are primarily gastrointestinal. Although it is generally not life threatening, it can make people quite ill. The illness is characterized by incapacitating diarrhea, nausea, vomiting, intestinal cramps, and the chills. With or without treatment, most people recover within a few days. The primary toxins responsible for DSP, okadaic acid and its derivatives the dinophysistoxins, are produced by dinoflagellates in the genus *Dinophysis*. Although DSP has been reported worldwide, the most highly affected areas have been in Europe and Japan. While DSP may not seem very serious, recent evidence suggests that prolonged exposure to DSP toxins may promote digestive cancers.

Neurotoxic shellfish poisoning (NSP) is primarily a neurological illness, with a rapid onset of symptoms. It is not life threatening. Its symptoms are similar to both PSP and CFP, but milder. The symptoms of NSP include numbness in the mouth, tingling in the mouth, hands and feet, poor coordination, and gastrointestinal upset. As with ciguatera poisoning, paradoxical temperature sensations may also occur (e.g. cold feels hot). NSP has only been reported in the southeast United States, eastern Mexico, and New Zealand.[33] The toxin responsible for NSP is brevetoxin, a neurotoxin produced by the dinoflagellate *Karenia brevis* (formerly known as *Gymnodinium breve*). Harmful algal blooms produced by *K. brevis* also cause massive fish kills. This toxin also is capable of becoming airborne. Wave action can break apart *K. brevis* cells and release the toxin in aerosol particles. Respiratory problems can result when the airborne toxins are inhaled.

Amnesic shellfish poisoning (ASP) is named for one symptom experienced by seriously ill victims: persistent short-term memory loss. People suffering from ASP typically experience both gastrointestinal and neurological disturbances. The most frequently reported symptom is vomiting, followed by abdominal cramps, severe headaches, short-term memory loss, and diarrhea. The symptoms are generally milder than seen for people experiencing PSP. ASP poses the greatest health risk for the elderly and patients with pre-existing medical conditions.

Of all the shellfish poisonings considered here, ASP was the most recently discovered. It was identified in 1987, in Prince Edward Island, Canada, when 156 people became ill after eating blue mussels.[34] It can be life threatening. Four people died in the Prince Edward Island outbreak. Although discovered in Canada, ASP is a continuing problem along the West Coast of the United States. ASP results from the production of domoic acid by diatoms in the genus *Pseudo-nitzschia*. Filter-feeding shellfish concentrate domoic acid, as they feed. The toxin can be present in shellfish tissues outside of their digestive organs. As a result, consuming shellfish after removing their guts is unwise. Analyses of crabs and fish tissues collected during domoic acid outbreaks revealed that the toxin was present only in the viscera, but not in muscular tissues. Therefore, crabs and fish appear to be safe to eat during domoic acid alerts, if their viscera are removed. This information was provided to the public in Oregon, several years ago at the height of a domoic acid event. It seemed to confuse the public, more than calm their fears. One principal reason was that few people knew exactly what "viscera" meant. People can't remove viscera without knowing what they look like. The term "viscera" is of Latin origins, the plural of viscus: an internal organ, in the body cavity. Viscera consist of the innards (the vital organs, the guts, the entrails) that are typically removed when cleaning a fish, or crab.

Fish poisoning

Ciguatera fish poisoning (CFP) is the most common type of poisoning that results from eating fish. The term "ciguatera" was first used in the Spanish Antilles, in the eighteenth century. It described an intoxication that resulted from the consumption of a marine snail, locally called "cigua" in Cuba. However, fish poisonings have occurred since ancient times. They were noted as early as 800 B.C., in Homer's Odyssey. And during his reign, Alexander the Great (356-323 B.C.) forbid his soldiers to eat fish to avoid the illness that might compromise his conquests.[35]

Ciguatera results from ciguatoxin and maitotoxin, biotoxins produced by the dinoflagellate *Gambierdiscus toxicus* that attaches to benthic (bottom) surfaces in many tropical and sub-tropical coral reef ecosystems. The toxins are ingested by herbivorous reef species as they graze, and are then passed up the food chain as these fish and invertebrates are ingested by carnivorous fish. When people eat the carnivorous fish, the biotoxin is ingested as well.

The onset of ciguatera symptoms ranges from minutes to hours after eating biotoxic fish. Conditions symptomatic of CFP can best be described as a complex of disruptions to the digestive, neurological, and cardiovascular systems.[36] The most common symptoms of ciguatera poisoning are a tingling and numbness around the mouth and throat, spreading to the hands and feet. Itching and paradoxical sensory perceptions also occur, such as cold feeling hot, and a shower feeling like electrical shock. Muscular aches and weakness also frequently occur. Disturbances to the digestive system include diarrhea most frequently, followed by vomiting, and nausea. Blood pressure may drop, and the heart beat may slow. Most of the time, the symptoms disappear within a few days. However, some people experience a recurrence of neurological disturbances for months, or (rarely) years.

In short, ciguatera can be very nasty, and should be avoided, if possible. As ciguatera only occurs in tropical and subtropical fish, whose food chain includes benthic-feeding, reef species, protecting ourselves against ciguatera can be accomplished in 2 ways, without eliminating fish from our diet. First, we can safely eat fish from temperate or higher latitudes. Salmon, halibut, lingcod, striped bass, and pollock are good examples of cold-water fish. Secondly, we can safely eat tropical fish that feed in open water, and not in close association with reefs. This is a little more complicated, because we need to understand the lifestyle of various fish species. Perhaps an easier approach is to remember which species are safe, and which species to avoid.

For people living at high latitudes, ciguatera poses the greatest risk for them when they travel to tropical and subtropical regions, especially when they catch their own fish. As ciguatera toxicity varies by locality, fish species, and fish size, local authorities can provide reliable advice on CFP toxicity in local fish.

The Centers for Disease Control and Prevention (known as the CDC) has listed the following as species known to carry ciguatoxins: barracuda, black grouper, yellowfin grouper, blackfin snapper, cubera snapper, dog snapper, hogfish, greater amberjack, horse-eye jack, and king mackerel.[37] The Center for Food Safety and Applied Nutrition lists barracuda, amberjack, grouper, snapper, jacks, wrasse, surgeonfish, eels, and parrotfish as the marine fish most frequently responsible for ciguatera poisonings.[38] However, the World Health Organization (WHO) declines to offer a list of fish which carry ciguatoxin, because they view the presence of the toxin as a local phenomenon, that is virtually unknown in some reefs. Instead they recommend obtaining advice locally.[39] If possible, ask experienced, local fishermen which areas to avoid and which fish are unsafe to eat. In the absence of specific information, avoid eating large, predatory reef fishes as they pose a greater health risk than small ones, due to their bioaccumulation of ciguatoxins.

Tropical and subtropical fish that are not likely to carry ciguatera toxins include mahi-mahi (also known as dolphinfish, and dorado), marlin, pomfret, spearfish, swordfish, tunas, and wahoo. However, some of these species are large predatory fish, and as a result, health advisories recommend limiting the consumption of large tuna, marlin, swordfish, and wahoo to 3 or fewer meals per month.

Fugu fish poisoning (FFP) is a very serious, often fatal type of poisoning, that results from eating fish in the order Tetraodontidea, which includes porcupine fish and puffer fish (known as fugu, in Japan). These fish are among the most poisonous marine life due to the biotoxin, tetrodotoxin, that is produced by marine bacteria that colonize mucosal layers in the skin and guts of infected puffers.[40] Tetrodotoxin, a powerful neurotoxin, is contained in the liver, gonads, intestines, and skin of fish in this group. This toxin is much more powerful than cyanide, and may cause death in 50 to 60 percent of people who are poisoned by it. Also known as puffer fish poisoning, FFP, is a rarer type of fish poisoning than ciguatera, but is potentially more dangerous. Most often, FFP results from the consumption of puffer fish from the Indo-Pacific region. The onset of fugu poisoning symptoms is rapid, generally occurring within 15 minutes to 2 hours after ingesting the toxin. Numbness and tingling around the mouth, face and extremities occur first, followed by nausea, vomiting, diarrhea, and abdominal pain. Weakness, depressed breathing, slurring of speech, and paralysis follow. Metabolism may be slowed to the point where a pulse and breathing are not detectable. Rarely, in Japan, people who are thought to have died from fugu poisoning recover completely from their apparently lifeless state.[41,42]

Only a portion of a puffer fish is edible, and preparing the fish without contaminating the edible portion is extremely difficult. In Japan, fugu chefs are specially trained and licensed. The training includes written examinations, and practical tests that include preparing fugu then eating it. Insuring that fugu consumption is as safe as possible requires following strict rules that cover cleanliness, storage, and preparation techniques, as well as careful documentation of the amount of fish handled and records of the toxic organs.

Many people might wonder why anyone would bother eating fugu if the potential health risks are so high. In Japan, fugu restaurants are marked by lanterns made from the dried skin of inflated puffer fish, and painted signs of puffer fish. They serve very expensive, multi-course meals, with each course featuring a particular presentation of fugu (including skin, flesh, and some organs) to discerning clientele. Diners claim that they appreciate the delicate flavor of fugu. Others claim that fugu testes in warm sake have aphrodisiac properties. But this is clear: when eating carefully prepared fugu, a diner experiences a tingling and numbing sensation in the mouth. The thrill of cheating death by surviving a "life-threatening" meal may be its greatest appeal. Fugu poisoning can be easily avoided by not eating puffer fish. And for those who need to experience the thrill, don't even consider eating fugu that is prepared at home. In Japan, most fugu fatalities result from home preparations of sport-caught puffer fish for two reasons: 1) lack of knowledge and skill to safely prepare it, and 2) wild-caught puffer fish contain significantly more tetrototoxin than the pen-raised puffer fish that are generally served in restaurants.

One additional way to protect ourselves against biotoxins is to avoid eating fish or shellfish that are sold as bait. They do not meet the same health-safety standards as seafood that is designated for human consumption. Buy seafood from reputable sources. For those who harvest their own seafood: pay attention to pertinent health advisories.

Over the last few decades, our collective concern over the state of the environment has been heightened. Print, audio, video, and electronic media have perceived this concern, and focused our attention through enhanced reporting of harmful algal blooms. Regardless of whether a bloom is called a red tide or not, it seems that toxic algal blooms have increased in frequency fairly recently. Otherwise, the general public's awareness of them has increased. This has led to many questions. Chief among these: Are we responsible for red tides? And do human activities increase the frequency and/or intensity of red tides?

Red tides occur naturally, regardless of human activity. Some oceanic regions that are severely affected by toxic algal blooms are rather remote and pristine. In the United States, Maine and Alaska, two states with relatively little industrial development and relatively small populations, both are severely affected by blooms of the algae that results in Paralytic Shellfish Poisoning. In both cases, the formation of these blooms can be attributed to specific conditions of physical geography, oceanography, and meteorology.[43] The rocky coastal topography of Maine and Alaska protrudes into cold water, and the continental shelf break is fairly close to shore. These physical conditions appear to promote the growth of massive populations of the toxic dinoflagellates, deep below the ocean surface. During periods when the ocean is thermally stratified, the blooms persist (usually from about mid-spring to mid-fall). However, winds can drive the surface waters offshore. When this happens the subsurface waters (and the blooms they contain) are forced into shallow inshore zones, where the toxic dinoflagellates are ingested by shellfish. Harmful algal blooms and their resulting shellfish intoxications are definitely naturally occurring phenomena.

While red tides have been around a long time, most likely pre-dating the existence of man, that doesn't mean toxic algal blooms are not affected by man's activities and the pollution that these activities generate. In some cases, increased pollution has been correlated with increased dinoflagellate blooms. Enriching a marine site with organic material and nutrients fuels algae blooms, but it is unclear how or why this pollution would favor toxic dinoflagellates over other phytoplankton species.

Dinoflagellates spend the winter in a dormant stage, a benthic resting cyst. These cysts re-seed the population, seasonally. However, some human activities such as dredging can result in the dispersal of the cysts and future toxic blooms to new locations.

When they are introduced to new areas, cysts could establish new, toxic dinoflagellate populations.

Is any seafood really safe to eat?

Rigorous health and sanitation standards have been established for seafood that is sold commercially in developed countries. In the United States, federal, state, and industry representatives voluntarily established the National Shellfish Sanitation Program (NSSP), a cooperative program that sets standards to ensure the safety of molluscan shellfish (a group of related invertebrates that generally have one or two hard shells, e.g. clams, oysters, conchs). States are responsible for enacting and enforcing laws and regulations relating to the harvest and processing of shellfish under sanitary conditions. The Federal component of this program, the (FDA), conducts annual reviews of each state program for compliance to NSSP standards. The shellfish industry contributes to shellfish safety by obtaining shellfish from safe sources, using sanitary processing plants, tagging and certifying shellfish, and keeping detailed records to document each package of shellfish. Water quality is monitored according to strict criteria. Shellfish is checked at the processor level, and at the market level.

In the US, the FDA has been given authority as the national agency responsible for public protection and regulation of seafood. Within processing plants, FDA inspectors consider hygiene and product safety. At FDA labs, samples are tested for contamination, spoilage, pathogens, additives, food dyes, drugs, and marine toxins. The FDA has the authority to detain or seize imported shipments if seafood is misbranded or adulterated. It also has the authority to issue health certifications that must accompany seafood that is exported to the European Union.

Seafood that is sold in markets or restaurants should be considered safe to eat and virtually free of the toxins described above. However, there are potential risks associated with gathering your own seafood, or eating fish or shellfish that has been gathered by friends or relatives. It's OK to ask where seafood came from, and decline to eat it if you don't like the answer you receive. Remember, cooking will not rid seafood of these toxins (nor will freezing, drying, pickling, salting, or smoking). And, unfortunately, the toxicity cannot be determined by the smell or appearance of seafood.

Whenever possible, shellfish such as crabs, lobster, clams, mussels, and oysters, should be cooked alive. The reason for this is that once shellfish die, digestive enzymes begin attacking their flesh. This decomposition results in an off flavor. Cooking live shellfish will stop this process, and yield the best flavor. Similarly, fish should be gutted as soon after catching as possible to minimize any post-mortem effects of digestive enzymes. Large fish should be bled by cutting the gills, when possible, to preserve the freshest taste and minimize oxidation of oils.

From a risk/benefit analysis, clearly the health benefits of consuming seafood outweigh the risks. However, the greatest health benefits will result when attention is paid to minimize avoidable risks. Before gathering shellfish, contact local or regional fish and wildlife facilities, or marine labs, or inquire at local bait shops to figure out the status of shellfish beds. If possible, call a shellfish hotline. Most coastal states maintain shellfish phone hotlines and/or websites. These are administered by a variety of state departments including Agriculture, Agriculture and Consumer Services, the Environment, Environment and Natural Resources, Environmental Conservation, Environmental Management, Environmental Protection, Environmental Quality, Environmental Services, Fish and Game, Health, Health and Environmental Control, Marine Resources, Public Health, and State Health Services. In some states, counties or cities maintain their own shellfish hotlines.

Please purchase recreational shellfish licenses wherever they are required. The money generated by license sales supports the regular testing of shellfish for toxins.

What's the truth behind shellfish myths?

The old adage that (in the Northern Hemisphere) it is safe to eat bivalve (two shelled) molluscs, such as clams, mussels, and oysters, in months with an "R" in the name (i.e., January, February, March, April, September, October, November, and December) is a fallacy with regard to biotoxins. Shellfish intoxications can occur during "R" months. In fact, this old rule pre-dates refrigeration, when spoilage was an extremely serious problem. The "R" months represent the cooler months, with lower water temperatures and lower bacterial counts. As a result, following the "R" rule during the pre-refrigeration era would have minimized the chances of people getting ill from eating raw clams or oysters, as a result of spoilage. On the other hand, as water temperatures rise, during the months that lack an "R" (May, June, July, and August), bivalve shellfish spawn. Spawning results in chemical changes in their tissues (for example, in glycogen content) that may seem to water down their flavor. While these chemical changes are not harmful, they may be another reason that some people avoided eating clams and oysters from May through August.

Advances in commercial harvesting procedures and refrigeration have generally enabled consumers to safely eat bivalves any time of year. Before using this generalization as a "free pass" to eat raw oysters in the summer, consider this: during warm months, bacterial counts naturally rise in marine environments. Shellfish feed by filtering everything out of the water including good and bad bacteria. Even legally harvested shellfish from non-polluted water can be contaminated after filtering pathogenic bacteria that inhabit their marine environment. Although these bacteria do not alter the taste, smell, or appearance of the shellfish, they can induce gastrointestinal distress in people who eat raw oysters. One pathogenic species that causes sporadic problems during warm months, especially in oysters along the Gulf of Mexico, is *Vibrio vulnificus*, a member of the bacterial family Vibrionaceae, which also includes the bacteria responsible for cholera (a severe diarrheal disease). Vigorous, healthy people are not likely to develop a serious *Vibrio* infection after eating contaminated raw or inadequately-cooked oysters. However, people with chronic health conditions, suffering from cancer, diabetes, or liver disease, or with impaired immune systems, can become extremely ill after eating contaminated raw oysters. Symptoms include gastrointestinal distress (diarrhea, nausea, stomach pains, and vomiting), as well as chills, fever, rash, and shock. *Vibrio* infections are fairly rare, but can result in septicemia, a life-threatening blood infection, in individuals who are medically at high risk. About half of *Vibrio* infections that move into the blood stream are fatal. Although cooking shellfish cannot negate the harmful effects of biotoxins, cooking does kill pathogenic bacteria and viruses. To minimize the chances of getting ill from bacteria in or on raw shellfish, we should only eat thoroughly cooked clams and oysters. For those who opt to eat raw shellfish: know its source, and make sure that it has been properly refrigerated after harvesting, because warm temperatures beget large increases in bacterial counts in shellfish.

Despite popular myths and anecdotal accounts, there is very little scientific evidence that pouring on hot sauce while eating raw or undercooked shellfish will kill all the pathogenic viruses and bacteria that may occur in raw clams and oysters. In 1993, scientists from Louisiana State University Medical Center reported putting *Vibrio* and other bacteria in test tubes, and then adding various components of hot sauce.[44] Straight Louisiana hot sauce killed all the bacteria. Even a six percent dilution of hot sauce killed all the bacteria within five minutes. Horseradish and lemon juice also killed bacteria, but were not as effective as hot sauce. While this may seem to be convincing evidence that hot sauce kills bacteria, and another "free pass" to consume raw oysters, we should remember that bacteria deep inside oyster tissues would not be as openly exposed to the destructive properties of hot sauce as naked bacteria inside a test tube. The digestive tract doesn't function like a test tube. If we swallow an oyster whole, bacteria deep inside the oyster would be pretty insulated from whatever effect hot sauce may have on the surface of the oyster. This idea was confirmed in 1995, by scientists from the University of North Carolina at Charlotte, who poured either Tabasco sauce or a horseradish-based cocktail sauce on freshly shucked oysters, and let them sit in the half shell.[45] After 10 minutes, the number of *Vibrio* cells on the surface and inside the oysters were counted. Tabasco sauce effectively reduced the number of *Vibrio* cells on the surface of the oyster, however, neither sauce effectively reduced the number of *Vibrio* cells inside the oyster.

As far as the myth that drinking alcohol while eating raw or undercooked shellfish offers protection against pathogenic bacteria: don't count on it. Although alcohol has been used as both a preservative and disinfectant for years, and high concentrations of alcohol will kill virtually all living things, who amongst us would consider drinking a glass of disinfectant-strength alcohol with each dozen raw oysters? Besides, abundant evidence suggests that excessive consumption of alcohol contributes to liver disease, and people with liver disease are extremely vulnerable to *Vibrio* infections. In the US, the typical shellfish-related *Vibrio* case can be described as a middle-aged, white man, who drinks heavily, and is either unaware of, or disregards, his medical risk factors. Rather than assuming that hot sauce or alcohol might afford us any significant protection from bacterial-laden shellfish, a wiser approach would be to eat only completely cooked shellfish. Thorough cooking kills both viral and bacterial pathogens.

The FDA provides the following guidelines for thorough cooking of oysters. Oyster shells should be closed. Discard any oysters that have opened their shells prior to cooking. Use small pots, in order to insure uniform cooking. If boiling, once oysters have opened their shells, boil for an additional 3 to 5 minutes. If steaming, place oysters in steamer over boiling water, and steam for 4 to 9 minutes. Discard any oysters that do not open during cooking. Shucked oysters should be cooked as follows: Boil or simmer for 3 minutes, or until the edges curl. Deep fry in 375°F (190°C) oil for at least 3 minutes. Bake at 450°F (230°C) for 10 minutes, or broil for 3 minutes.[46]

As to whether oysters are aphrodisiacs, there is little scientific evidence to support this claim. Most arguments have centered on the high zinc content in oysters, because zinc is a constituent of sperm, and zinc deficiency can lead to male impotence. However in 2005, a team of American and Italian scientists discovered that oysters are rich in two relatively rare amino acids that trigger the production of sex hormones: testosterone in males, and progesterone in females.[47] These amino acids, D-aspartic acid and N-methyl-D-aspartate are at their highest levels in the spring, when bivalves are spawning. To get the maximum effect of these amino acids, oysters should be eaten raw because cooking diminishes them. But of course, eating raw oysters is not risk free, in terms of human health.

Another shellfish myth which persists, despite evidence to the contrary, is that shellfish contain high levels of cholesterol, and should not be eaten by people who are trying to minimize cholesterol in their diets. Shellfish contain sterols, a class of chemical compounds that includes, but is not limited to, cholesterol. Early chemical analyses of cholesterol content were inaccurate, and not sophisticated enough to differentiate between cholesterol and other sterols. Rather, they measured total sterol levels instead of simply cholesterol. With the advent of more refined chemical analyses, the cholesterol level in shellfish has generally been shown to be less than levels in comparable portions of meat or poultry. Clams, oysters, mussels, scallops, crab, and lobster all have low levels of cholesterol. Shrimp have the highest levels of cholesterol of shellfish, but contain much less fat (especially saturated fat) than meat or poultry. The saturated fat content of a diet may largely be responsible for increasing blood cholesterol levels. Although shellfish generally contain low levels of cholesterol and are low fat, the healthfulness of eating shellfish can easily be compromised by cooking methods. For the most healthful diet, avoid dredging shellfish in butter, deep frying, or smothering it with mayonnaise or cheese.

Is a seafood diet good for the planet?

Perhaps one of the most important ecological issues facing people who want to add more seafood to their diet is knowing which varieties of seafood are harvested in a sustainable manner. Fisheries can only endure when seafood is captured, grown, or collected in a manner that is, in the long term, not detrimental to the overall health of seafood populations and marine ecosystems. We all need to know that what we eat for dinner tonight isn't going to compromise our ability to enjoy the same meal two years or twenty years from now.

More than half of the fisheries of the world are categorized as either "fully exploited" or "over exploited." Essentially, too many fishermen in too many boats are fishing for too few fish. When this happens, fish populations are overfished and catches decline. In response to declining catches, boats often fish longer and try new sites or new methods. Increasing fishing efforts and using technologically advanced gear to locate and haul in catches both exacerbate the problem. Short-term increases in catch may be achieved, but eventually they result in even more devastating drops in fish populations.

At the present time, it is clear that marine fishery resources are not limitless, as was once believed. For hundreds of years, man has fished various stocks to depletion and then moved on to repeat the process in a different location, or with a different species. The size of fish that are being caught has become progressively smaller. The important giant breeder fish have been removed from many stocks. Fishers have overfished the top predators and then continued fishing down the food web. It is clear that with the world population now over seven billion, we cannot expect business-as-usual fisheries to be able to provide enough seafood to satisfy world demand. Most fisheries are not sustainable. Yet, sustainable fisheries are the only hope we have to be able to enjoy seafood now and in the future without endangering seafood populations. In order to understand the importance of eating sustainable seafood in our diet, we need to consider the journey from the sea that ends as food on our plates. Let's start with fishing.

How has fishing evolved?

Utilizing marine resources has been an important part of human culture for thousands of years. This has been established through the recovery of artifacts and faunal remains at archaeological sites around the world. Early humans engaged in marine foraging for shellfish. They gathered mussels and sea snails at Pinnacle Point, in coastal South Africa, 164,000 years ago.[48] The oldest evidence of sophisticated fishing technology was also found in Africa. At Katanda, in the Upper Semliki Valley, Zaire, barbed bone points dated at 90,000 years ago appear to have been used to harpoon catfish, during the Middle Stone Age.[49] More recently, through the Later Stone Age or Upper Paleolithic (circa 40,000 to 10,000 years ago), at various sites, such as the Dordogne Valley, France, fishing technology incrementally gained sophistication. During this period, early, anatomically-modern humans fished using barbed bone and antler points that were crafted into spears and harpoons, nets made of vines, and traps. Archeological sites at the eastern end of East Timor, in the Indonesian archipelago, recently provided evidence that early modern humans consumed pelagic marine fishes (such as small tunas) 42,000 years ago. Based on both the weight and number of bones present, fish were the most important prey. And from 42,000 to 38,000 years ago, pelagic and inshore fishes were essentially equally represented at these sites. Although no evidence was found to document fishing technology during the earliest period at this site, a broken shell fishhook was dated at 23,000 to 16,000 years ago. This represents the earliest conclusive evidence of hand-crafted fishhooks in the world.[50] Not surprisingly, fishing methods and gear have changed dramatically through the millennia, yet the sea has been a vital source of food for millions of people, around the world, for a very long time.

For centuries, the right to fish in waters that were close to shore and easily accessible to local fishers was mainly controlled at the local level. And more distant fisheries were exploited internationally on an un-regulated, open-access basis. In the open seas, fishers could catch whatever they wanted, whenever and wherever they wanted. This laissez-faire system worked for quite a while. However, as the earth's population increased dramatically, and fishery resources began getting scarcer, this free-for-all strategy created conflict, and coastal countries realized that access to the fisheries that were close to their countries, and vital to their economies needed to be regulated. Midway through the twentieth century, several countries began unilaterally claiming jurisdictional authority beyond their territorial waters, extending out to the continental shelf. Throughout the final third of the twentieth century, serious international dialogue on how to divide up oceanic resources, control pollution, and define navigational rights commenced and evolved into the Law of the Sea. Basically, the Law of the Sea extended each coastal nation's sovereignty, including the regulation, management, and economic control of marine natural resources such as fishing grounds, out to a 200-nautical-mile (370-km) boundary. Once ratified by the requisite number of countries, in 1994, this became international law as part of the United Nations Convention on the Law of the Sea. Although the United States abides by many of the articles of the convention, the US Senate has neither ratified nor acceded to it.[51] The 200-nautical-mile exclusive economic zones (EEZs) of coastal nations have become legitimately recognized around the world. Countries with extensive seacoasts, such as the United States, benefitted from this arrangement, and landlocked countries had to negotiate for access to prime fishing grounds. Despite the fact that the establishment of EEZs was intended to benefit the conservation of fishery resources, in some cases, such as the Canadian and American NW Atlantic cod fishery, due to economic forces, it merely swapped over-exploitation of fisheries by foreign fleets with over-exploitation of fisheries by over-capitalized national fleets.

Fisheries are composed of several types of fishers. Traditional fisheries that employ skilled, but non-industrial fishers are considered artisanal. Historically, around the world artisanal fisheries predominated. Today they continue to be very important principally in developing countries. Artisanal fishing, also known as small-scale or subsistence fishing, is generally family or community based. Even though definitions vary from nation to nation, in developing countries artisanal fisheries often rely on small, open boats, and simple gear, which sometimes includes a small outboard motor. The artisanal fishery is very low-cost, low-tech, and labor intensive, but nevertheless it makes significant contributions to the well-being of local communities, both nutritionally and economically.

Unfortunately, due to the low-tech nature of the artisanal fishery, a large portion of the catch may be lost before it can be sold or consumed. In many cases, particularly in impoverished tropical regions, losses are principally due to spoilage when catches are neither refrigerated nor processed in a timely fashion. Between the time a fish is captured and consumed, it must be transported, processed, stored, and marketed. If the fishing is good, fishing trips are relatively short, but the catch may languish at processing plants or markets that can't handle the bounty. In some cases, when too many fish are landed, they are simply dumped. If fishing is poor, fishing trips are often extended and the catch may languish in the boat, spending up to 12 hours exposed to the sun. Although small-scale fisheries can be productive, their post-harvest losses need to be minimized. As countries develop economically, their fish-processing infrastructure modernizes, and fishing shifts towards bigger fishing boats equipped with more modern gear. A large portion of the world's fishing fleet consists of medium-sized fishing boats, equipped with fairly modern gear.

Although fishing has existed for tens of thousands of years, some fishers still catch fish using the most primitive methods: with their bare hands, from the shore. However, most fishermen utilize much more advanced fishing techniques and gear in pursuit of their catch. The earliest advances in fishing gear were spears, hooks, nets, and traps. Small crafts that were pushed or pulled along by poles, paddles, or oars, and boats that were propelled by the wind in their sails enabled fishers to move their fishing expeditions away from the shore. Over time, small sailboats were replaced by wind-driven sailing ships, which were in turn replaced by steam-driven ships. Wood provided the earliest fuel for the steam-driven, piston engines, followed by coal, then diesel. Eventually, more efficient, diesel engines replaced steam engines. With each advance in ship propulsion came new fishing opportunities. Commercial fishing moved out of rivers and lakes, and away from the seashore. For hundreds of years, sailing ships enabled fishers to travel vast distances from their home port, in search of distant fishing grounds. Then, early in the twentieth century, steam engines began replacing sails, freeing fishers from fishing at the mercy of prevailing winds. However, the operation of steam engines on long voyages required carrying a lot of fuel, and fresh water. Diesel-electric engines provided propulsion, as well as an excellent power source to drive the winches and blocks that are used to haul in fish-laden nets and other gear. The mechanization of fishing was greatly facilitated by diesel-electric engines. Larger nets could be fished and retrieved.

Just as fishing vessels evolved, so did fishing gear. For example, early nets were hand made out of naturally occurring fibrous materials, such as flax, wool, and local grasses. As the size of fishing vessels and their power increased, nets were enlarged. Later, mechanical net making replaced the making of nets by hand. Over time, nets made of natural fibers were prone to rotting. However, by the mid-twentieth century, following World War II and the development of synthetic fibers such as nylon, natural fibers were replaced by rot-resistant synthetic fibers for net construction. Nets made with long-lasting, synthetic fibers were cheaper, easier to handle, and required less maintenance than natural-fiber nets. Additionally, monofilament nets appeared virtually invisible underwater, and as a result generally caught more fish than natural-fiber nets. Fishermen were drawn to synthetic-fiber nets for all of these reasons. Years later, marine biologists realized that the long-lasting nature of synthetic fibers had a down side as well, because lost nets that were constructed out of synthetic fibers continued to catch fish, as they floated aimlessly through the sea.

Technologically, fishing has advanced incredibly since the middle of the twentieth century. Many of the technologies that were developed initially for wartime military uses have been re-engineered to benefit fishermen in their quest to find fish. Onboard radar enables captains to monitor approaching weather as well as ship traffic. Global positioning systems (GPS) enable fishermen to record the exact location of reefs, seamounts, and other physical structures around which they find aggregations of fish species, and to mark where they have deployed fixed gear such as traps. Sonar has morphed into an echo-sounding tool to find fish stocks, and map sea-bottom terrain. While hydroacoustic surveys are used by fishery agencies to estimate fish populations, commercial fishermen are using their own hydroacoustic transducers to locate fish, especially schooling species. Even small, sport-fishing boats are often outfitted with both fish finders and GPS devices.

From an economic perspective, the pinnacle of development in fishing is achieved by expensive, high-tech industrial fleets of factory ships, that both catch and process fish. These vessels are very expensive to operate. And, in order to be profitable, they must work at or near capacity, 24 hours a day, hauling in huge catches of valuable fish or shellfish species. In some cases, their catches seem to virtually suck the life out of fishing grounds. When fishing in the waters off developing countries, industrial fleets can literally take food out of the mouths of hungry locals, as they out-fish native artisanal fishers. This problem is exacerbated when the governments of poor countries sell the fishing rights within their waters to foreign fleets, in exchange for hard currency to pay down their debts. Although many developing countries have productive fishing grounds in the waters that make up their EEZs (e.g. off Africa), these resources do not feed their hungry citizens, rather the catches within their EEZs are generally destined for high-priced export markets.

Globally, only two sources of seafood production exist. Seafood is either captured in the wild or harvested from an aquaculture facility. For centuries, capture fisheries produced the vast majority of our seafood, but in recent decades, production from aquaculture has grown rapidly. In either case, seafood resources need to be utilized in sustainable ways that do not permanently deplete or damage their populations. The resilience of fisheries should be considered in a generational time sense. We should not squander today's marine resources. Rather, we should insure that marine ecosystems are healthy and productive for our children and grandchildren as well as their children and grandchildren. These resources need to be managed wisely. Sustainable fisheries require cooperation, knowledge, and trust between the fishers and the managers. In order for fishery management to be successful, fishers must feel that the decisions being made are in their best interests, as well as those of the fish populations.

We should be cognizant that humans do not represent the only stake-holders in ocean fisheries. We are not the sole consumers of the seafood resources of the world. Although we may try to maximize catches of our favorite species, and we may try to manage marine resources to benefit ourselves, vast numbers of other species also depend on marine food webs to survive. Many of these species live beneath the ocean, and are rarely, if ever, seen by humans. These essentially anonymous consumers of seafood consist of thousands of species of fish and invertebrates, including giant squid that feed in the deep ocean on Patagonian and Antarctic toothfish. Although we share a greater emotional connection with the air-breathing denizens of the deep: sea turtles and marine mammals, especially dolphins, seals, sea otters, and whales; we rarely consider them potential competitors for some of our favorite seafood meals. Similarly, we are enamored of the grace and beauty of marine birds that live in the open ocean for much of the year, and come ashore only to nest, yet we rarely think about what sustains them. These species rely on the same, naturally fluctuating, marine seafood production as our fishing fleets. A truly sustainable fishery should support all of its stake-holders.

What makes a fishery sustainable?

In order for a capture fishery to be sustainable it should minimize the waste and destruction of marine resources and habitats. Sustainable fisheries allow for the indefinite use of fishery resources. Methods of catching fish or invertebrates should not destroy or cause long-term damage to the habitat in which they are used. Fisheries are sustained from year to year through recruitment, i.e., from new fish that are added to the exploitable stock each year, as fish grow large enough to be caught, or when fish migrate into a fishing ground. In a sustainable fishery, the catch should never exceed recruitment into the fishery, and the amount of fish harvested should not reduce the biomass from year to year.

Commercial fishing gears are divided into two types: active gear and passive gear.[52] Fishing vessels pull active gears through the water, to capture their target species. These gears actively chase the catch. Three basic kinds of active gear are trawls, dredges, and encircling nets. Trawls, wide-mouthed, tapering, bag-like nets, are used in mid-water and near the bottom to catch fish and invertebrates such as shrimp. Dredges, metal nets with metal frames, are dragged along the sea floor to catch benthic invertebrates such as scallops, oysters, and clams. Encircling nets, large nets that surround fish schools from the sides and the bottom, are fished by two boats, in open water for pelagic species. Passive gears are set in one place, or drift with currents, and the target species catch themselves in the gear, often as a result of being attracted to bait. Four basic kinds of passive gear are gillnets, longlines, traps and pots, and hook and line. Gillnets, large, net panels made of netting that fish can't easily see, hang vertically in the water, and drift with currents. Longlines utilize a large number of branch lines with baited hooks attached to a long, stationery mainline. Baited traps and pots that rest on the sea floor are used principally to catch crustaceans such as crabs and lobsters and a few fish species. Fishing via hook and line involves individual lines with baited hooks.

Bottom trawls are one type of fishing gear that has been extremely damaging to bottom habitats. Many commercially valuable fish species, known as ground fish or demersal fish, live close to the bottom of the water column and depend upon sea-floor habitats. One method of fishing for ground fish is to drag a trawl net across the sea floor. These bottom trawls frequently tear up complex sea-floor habitats that are important for many benthic and demersal species. Scallop dredges similarly cause serious damage to both physical and biotic components of benthic habitats. Long-lived, benthic species such as sponges, soft corals, and hard corals, that provide shelter and habitat for a variety of fish species and their prey, grow slowly and can take years, or even decades, to recover from unsustainable fishing practices such as using bottom trawls or scallop dredges.

In a sustainable fishery, the fishing gear should be selective enough to catch the desired species. The fishery should not inadvertently catch and then discard a large biomass of extraneous species. The gear should also avoid catching juvenile year classes, as juveniles represent the fishery production of the future. Seasons and geography are also factors that need to be considered, in order for a fishery to be sustainably harvested.

Sometimes, very specific timing contributes to whether a fishery is sustainable. For example, closing a fishery on known spawning grounds for a few days can greatly benefit certain fish populations. Marine fish often have very complicated life histories. In most cases, males and females release their gametes (eggs and sperm) into the water column. After spawning, the fertilized eggs essentially drift along with water currents as they develop. If spawning occurs at the wrong time or wrong place, developing eggs and larvae will be swept away from the resources that they need, and their survival will be endangered. In order to maximize the success of their offspring, some fish species have evolved reproductive behaviors that make them especially vulnerable to over-fishing. Groupers provide a good example of this. For thousands of years, groupers have spawned at very specific times and places. In the case of the Nassau grouper,

Epinephelus striatus, a large sea bass that lives in coral reef habitats in the tropical western Atlantic Ocean, Gulf of Mexico, and Caribbean Sea, for a few days around the full moon in winter, males and females swim as far as hundreds of miles/kilometers to spawn at historical spawning sites. Oceanographic conditions at these sites favor the widespread transport of developing embryos within highly productive currents. Although many generations of fishers have known when and where grouper spawned, and generally respected them, the spawning aggregations of thousands of large grouper unfortunately have been irresistible to some fishers looking for a big payday. Fairly recently, various governments and environmentalists have tried to stop or curtail fishing during these spawning aggregations, but the conservation efforts have occurred after most of the known spawning sites have disappeared. Around the world, grouper populations have been seriously overfished. Even though grouper continues to appear on many menus, basically it should not be considered an ecologically smart food choice, until stocks recover.

Another factor that complicates whether a fishery can be sustainable or not, is the life style or strategy of a species. Some species grow quickly, mature early, and reproduce abundantly during a rather short lifespan. They live opportunistically, and are able to take advantage of ephemeral resource richness. The life style of these species is tied to a high reproductive rate. Their populations tend to be quite large, but may experience swings in size from year to year, depending on local conditions. In contrast, another group of species grows slowly, reaches sexual maturity later in life, and produces fewer offspring, during a long lifespan. The life style of these species is determined by the biological carrying capacity of their environment, i.e., their population size is principally shaped by what the resources in their habitat can support. Their populations tend to be smaller, and relatively stable from year to year.

Which fisheries are sustainable?

Species that grow rapidly and have large populations are better suited for sustainable fishing than slow-growing species. Small pelagic fish, i.e., fish that spend virtually their entire lives in the water column, continuously swimming in immense schools, are examples of fast-growing species that can be sustainably fished. Although industrial fisheries target many of these fish for processing into fish meal and oil, some make fine human food as well. These include herring, sardines, and anchovies.

The herring family (Clupeidae) consists of over 200 fish species of great ecological and economic significance, including alewife, herring, menhaden, pilchard, shad, and sardine species. Typically, clupeids are small, silvery, streamlined fish that feed on plankton. Some herring species feed as primary consumers, eating phytoplankton, and others feed as secondary consumers, eating zooplankton. Their biomass fuels many marine food chains, including as bait in lobster and salmon fisheries. Various herring species comprise roughly 50 percent of the annual global production of capture fisheries. Commercially important species include Atlantic herring (*Clupea harengus*), Baltic herring (*Clupea harengus membras*), Pacific herring (*Clupea pallasii*), European pilchard (*Sardina pilchardus*), and Atlantic menhaden (*Brevoortia tyrannus*). Of these fish, only menhaden is not consumed as a human food. Herring is marketed in a variety of forms, including filleted, marinated, pickled, salted, smoked, and canned. Herring roe is considered a delicacy in Japan, and shad roe is considered a delicacy in the US. Norway, a country with a proud fishing tradition, serves herring in an incredible variety of preparations, at every meal, including breakfast buffets. Scandinavian smorgasbords can feature as many as 20 different preparations of herring. Even the British enjoy traditional breakfasts that feature kippers, i.e., salted, cold-smoked herring.

The term "sardine" is not used exclusively for one species of fish or even a single genus. Instead, around the world "sardine" generically refers to one of several small, soft-boned, oily fish species in the herring family, including herring, sprat, and pilchards. Although some Europeans consider young European pilchards as the only true sardines, 20 other species in the genera *Sardinops, Sardinella, Clupea, Sprattus, Hyperlophus, Nematalosa, Etrumeus, Ethmidium, Engraulis,* and *Opisthonema,* are also permitted to be marketed as sardines, by World Trade Organization agreement. Sardines are bony, and because of their small size, de-boning the fish isn't feasible. However, the canning process virtually dissolves the little bones, resulting in a healthful seafood choice that is high in calcium. Sardines are most frequently found canned in US markets. Occasionally, fillets or whole adult sardines are marketed.

As an extremely oily, very bony, smelly fish that deteriorates rapidly following capture, menhaden is generally considered unfit for human consumption. Despite this, menhaden has been called *The most important fish in the sea* by H. Bruce Franklin in his 2007 book,[53] due principally its role as a primary consumer that filters algae cells out of the plankton. Although the diet of most other fish species in the herring family consists principally of zooplankton, menhaden principally consume phytoplankton. In doing so, menhaden convert the primary production of plant cells (algae) into fish flesh that is readily eaten by a variety of predatory fish species, including striped bass *(Morone saxatilis)*, weakfish *(Cynoscion regalis)*, and bluefish *(Pomatomus saltatrix)*. By ingesting plant cells, menhaden are capable of cleaning up algal blooms, while at the same time producing a rich source of prey for higher trophic levels in marine food chains. In terms of ecological energetics, only about ten percent of the energy consumed by one trophic level is available for consumption by the next trophic level. Significantly less energy is available to each successive link in a food chain. And so, short food chains that involve menhaden can potentially make more high-quality energy available to piscivorous (fish-eating) predators than would long trophic chains that run through several levels of zooplankton predation. The high content of omega-3 fish oils in menhaden has made them a prized catch of the industrial (reduction) fishery that supplies processing plants. These factories render menhaden into fish oil (which is then further processed into various products including purified omega-3 fish oil supplements for humans), fish meal, animal feed, and/or fertilizer. In recent years, the demand for heart-healthy omega-3 oils has increased as they have been incorporated into an incredible assortment of food items, ranging from margarine and milk, to bread, cereal, orange juice, and eggs. Adding omega-3 fish oils to food products gives them added nutritive value, as well as higher prices. Omega-3 oils are also being incorporated into various beauty products, for skin and hair care. All these products are then marketed as more healthful than products without the supplements. However, as a result of adding omega-3s to a seemingly endless variety of products, as well as the increasing demand for fish meal, many menhaden stocks have been seriously overfished. Without access to menhaden, their preferred prey, populations of striped bass have starved, especially in Chesapeake Bay. Wild populations of many of the fish species that we like to eat would benefit if tighter restrictions were placed on the industrial fishery for menhaden. Additionally, bodies of water that are plagued by immense algal blooms could benefit from the ability of large populations of menhaden to graze down the blooms, and clear the water. Menhaden fill a couple of large ecological roles in marine ecosystems that cannot be easily replaced by another fish species. Hopefully, fishery managers will recognize this before menhaden populations are fished to the point of no return.

The anchovy family (Engraulidae) contains a number of species, however, most of the anchovies sold in the US are the European anchovy (*Engraulis encrasicolus*). Like sardines, anchovies are also small and bony. They are most frequently sold canned, although salt-packed (dried) anchovies can often be found in Italian-style delis.

Larger pelagic species that are fast-growing may also be sustainably fished. Examples include Atlantic mackerel (*Scomber scombrus*), and wahoo (*Acanthocybium solanderi*) both members of the tuna family (Scombridae). Of these two, Atlantic mackerel is the smaller, rarely exceeding 2.2 pounds (1 kg) in weight. Mackerel live up to 20 years, in cold and temperate shelf waters. It is an oily, richly-flavored fish, with grayish flesh, that whitens as it cooks.

Most of the US harvest of Atlantic mackerel is exported. Wahoo can be described as the long, slender, tropical cousin of mackerel. They commonly reach lengths of 60 inches (1.5 m), with a world record weight of 184 pounds (83.5 kg), although only living up to 9 years. It is a mild-flavored fish, with flesh that whitens as it cooks, and is significantly less oily than mackerel. Wahoo populations are found in tropical and subtropical waters of the Atlantic, Pacific, and Indian Oceans.

Another good-sized, pelagic fish that is caught in a sustainable fishery is mahi-mahi, also known as dorado and dolphinfish (*Coryphaena hippurus*), one of only two species in the family Coryphaenidae. Mahi-mahi is a fast-growing fish that can reach 6 feet (2 m) in length, despite living only 4 to 5 years. It is a lean, mild-flavored, tasty fish, that occurs in sub-tropical and tropical waters of the Atlantic, Pacific, and Indian Oceans.

One of the most abundant fishes in the north Pacific, walleye pollock (*Theragra chalcogramma*), a member of the cod family (Gadidae), supports the world's largest whitefish fishery. Prior to the development of sophisticated processing techniques, this relatively fast-growing, short-lived species was considered unmarketable, due to its soft flesh. Now, factory ships are able to process the fish at sea. While markets exist for pollock fillets and roe, much of the fish is processed into surimi (a flavorless fish puree) which is then flavored, textured, and shaped into a variety of products, including fish sticks and artificial crab meat (known as krab meat). Choosing to eat north Pacific pollock, in any form, makes sense ecologically, as this fishery has been certified as a best environmental choice by the Marine Stewardship Council (MSC). Their standards are the only internationally recognized method of assessing whether fisheries are well managed and sustainable.

Wild-caught Alaska salmon are considered sustainable. Five species of salmon are caught in the Alaskan fisheries in the north Pacific: Chinook or king (*Oncorhynchus tshawytscha*), coho or silver (*Oncorhynchus kisutch*), sockeye or red (*Oncorhynchus nerka*), chum or keta (*Oncorhynchus keta*), and pink or humpback salmon (*Oncorhynchus gorbuscha*). King salmon is the largest Pacific salmon species reaching up to 5 feet (1.5 m) in 8 years, but each of other species grows fairly rapidly, in 2 to 5 years. The color of the flesh varies by species, with red salmon having the darkest red meat. King and red salmon are both very rich in omega-3 fish oils. As a result, the two are the most highly prized salmon caught in US waters. All of these species are tasty. Pink salmon is the most frequently canned salmon. Chum salmon is leaner and less oily than other salmon species, and sells at lower prices. The MSC has certified wild-caught Alaska salmon as a best choice, and sustainable. For those who choose to fish for their own salmon: be sure to check with the local fish and game authority for any and all regulations. They can be quite complex, with specific dates and catch limits for each river system.

Although fast-growing, short-lived fish species typically are the most likely candidates to support sustainable fisheries, a few slow-growing, long-lived species maintain stable fisheries. One example, Pacific halibut (*Hippoglossus stenolepis*), the largest flatfish species, a member of the right-eyed flounder family (Pleuronectidae), lives up to 50 years, and can weigh as much as 500 to 700 pounds (225 to 315 kg). Yet the Pacific halibut fishery, which is carefully managed, both in terms of commercial and sport catch, by the International Pacific Halibut Commission, has been certified as a best environmental choice by the MSC.

Another smart, sustainable fish choice that is even longer-lived, is black cod, also known as sablefish (*Anoplopoma fimbria*). Despite being called "black cod," it is not a cod at all, instead it is a member of the sablefish family (Anoplopomatidae). Adult sablefish are found in deep water, from 650 to over 6000 feet (200 to 2000 m), in the coastal north Pacific from California to Japan. They can live over 90 years, and reach 4 feet (1.2 m) in length. However, they grow fairly rapidly when young, and mature at 5 to 7 years, which is relatively early for such a long-lived species. Their fishery is well managed, and they are caught using pots (fish traps) or bottom longlines, with relatively little impact on non-target species. This white-fleshed fish, with a high fat content, has a sweet flavor and appealing texture. Although it may be gaining market share in the US, sablefish is extremely popular in Japanese cuisine, and the vast majority of the world catch goes to Japanese markets. It is also in demand as a smoked fish in various cuisines, including as smoked sable in Jewish delis.

Farmed fishes such as catfish and tilapia that feed as herbivores are ecologically great fish choices for a very important reason: their feed is not based on fishmeal. Various catfish and tilapia species have been farmed around the world for quite some time. Tilapia farming dates back 4000 years, to ancient Egypt.[54] Tilapia comprise a large tribe of African freshwater fishes in the family Cichlidae. Hybrids of several north African tilapia species have been extensively farmed around the world. Catfish species are cultivated in Africa, Asia, and the US. Because both catfish and tilapia are reared in closed, inland ponds, where they are fed vegetarian feeds, these farm-raised fish do not deplete wild ocean fisheries of bait fish.

Other sustainable seafood choices include invertebrate species such as crabs and bivalve molluscs: clams, mussels, oysters, and scallops. Of the crab fisheries, Dungeness crab along the Pacific coast and stone crab in the Gulf of Mexico and Atlantic coasts are the most sustainable. All of the bivalves that are grown out in farmed conditions are superb, sustainable seafood choices. Harvesting farmed shellfish that have either been outplanted in estuaries or grown in bags, racks, or on lines causes far less damage to benthic ecosystems than the dredges that are dragged across the seafloor to harvest wild-caught clams and scallops. And all bivalves feed by filtering algae out of the water column, so they don't require high-protein feeds derived from fishmeal. Rather, they can actually improve water quality by cleaning up algal blooms.

Fast-growing, abundant, short-lived, squid species should also be considered reasonable food choices, ecologically. Several small species of squid are harvested around the world. However, many of their populations aren't clearly understood, so squid aren't given a "best choice" designation. Additionally, squid provide abundant prey for many marine food webs.

To help you make sustainable choices, several marine conservation programs have developed dining guidelines: lists of ecologically correct (and ocean friendly) seafood choices. The Monterey Bay Aquarium encourages consumers to make informed choices and provides seafood recommendations through their outstanding Seafood Watch which can be found on their website.[55] The Seafood Watch can be browsed or searched by species. Regional seafood guides are available, listing the best seafood choices, reasonable alternatives, and seafood to avoid. Pocket-sized, printable copies of the Seafood Watch guide can be downloaded, or the guides may be downloaded as smartphone applications. Guides are provided for the northeast, southeast, central, southwest, and west coast of the US, Hawaii, and for sushi. A national Seafood Watch guide, and the west coast guide are available in Spanish. Another incredibly useful guide is the Blue Ocean Institute Seafood Guide which is posted on their website.[56] This guide evaluates seafood in terms of the abundance of fishery stocks, and the sustainability of fishing or farming techniques. It also identifies which species may present health risks due to mercury or PCB levels. The Environmental Defense Fund (EDF) also provides a printable buying guide to seafood, the Pocket Seafood Selector,[57] and a Pocket Sushi Selector[58] on their website. The EDF guides lists eco-best and eco-worst seafood choices. Additionally, they indicates which seafood choices are good for us because they are high in omega-3 fatty acids and low in contaminants or bad for us because they tend to be high in mercury and/or PCBs. In these guides, seafood is generally designated from green to yellow to red along an ecological goodness scale. Green defines relatively abundant, well-managed seafood that you should enjoy. Yellow defines seafood to be careful about eating. And red defines seafood to avoid, because of overfishing, or unsustainable fishing practices. The Audubon Society's Seafood Wallet Cards are no longer available for downloading, but in 2000, Audubon published the *Seafood Lover's Almanac*,[59] and they continue to support sustainable fishing and ocean conservation. The Almanac provided consumers with an entertaining blend of the life history and ecological status of seafood species, as well as recipes. The Smithsonian has also contributed to promoting sustainable fisheries by publishing their informative *One fish, two fish, crawfish, bluefish: the Smithsonian Sustainable Seafood Cookbook*[60] in 2003.

Books provide a wealth of additional information that could further empower you in making smart seafood choices. Several recent works that have dealt with the issue of sustainable seafood were written for a general audience. These include *The end of the line: how overfishing is changing the world and what we eat*,[61] *Bottomfeeder: how to eat ethically in a world of vanishing seafood*,[62] *Sushi: a guide to saving the oceans one bite at a time*,[63] and *Four fish: the future of the last wild food*.[64]

To facilitate shoppers making sustainable seafood selections, some grocery chains and fish markets have adopted the Monterey Bay Aquarium and Blue Ocean Institute's guidelines, and are posting Seafood Watch color-code ratings that reflect the ecological impact of each fishery. Austin, Texas-based Whole Foods Market, Inc., was the first national chain to display these ratings. This enables customers to know the environmental impact of the seafood that they purchase. The easy to remember color codes parallel the colors of a stoplight, green-coded seafood is good, go ahead and eat it. Be cautious (or go slow) in your consumption of yellow-coded seafood, and stop eating seafood that is red-coded. And as of Earth Day 2012, Whole Foods no longer carries un-sustainable, red-coded seafood.[65]

Which fisheries are not sustainable?

Populations of long-lived fish species grow slowly, mature late in life, and have relatively low reproductive rates. Each of these characteristics makes a species unable to withstand sustained heavy fishing, as is clear from their declining catches. Their populations are very slow to recover from overfishing. Examples of fisheries that are un-sustainable for this reason include orange roughy, Patagonian toothfish, and sharks.

Orange roughy (*Hoplostethus atlanticus*), a member of the slimehead family (Trachichthyidae), is a bright orange, deepwater fish with rough scales. This species is caught principally in the Southern Ocean using bottom trawls, a type of fishing gear that damages vulnerable, deep-sea-floor ecosystems that are slow to recover. Orange roughy grow slowly, live to over 100 years old, and don't reach sexual maturity until between 22 and 40 years old.[66] The ease with which this species can be captured, its low productivity, and its high value are inherent factors that make the fishery unsustainable.[67] The firm, white flesh of this oily fish has a delicate flavor that makes it popular around the world. Boneless fillets of orange roughy freeze well, and the US represents its biggest market. Until about 30 years ago, orange roughy populations survived deep beyond the reach of humans. But modern technology enabled fishers to locate and catch the species. Orange roughy gather in large numbers around seamounts to spawn and feed. Their fishery has been likened to a lucrative, but brief, gold rush. Because they live longer, and breed later in life than most other fish species, orange roughy populations are less productive and take longer to recover from heavy fishing pressure than other fish. Many of their fishery hot spots have demonstrated a boom, followed within a few years by a bust, from which they may never recover. One sobering aspect of the orange roughy fishery was that early catches at seamount hot spots yielded such tremendous volumes of fish that handling and processing facilities were overwhelmed, and much of the catch was simply dumped.[68] While

New Zealand and Australia currently have fairly good management programs for orange roughy fisheries, outside of their waters, other fisheries are poorly regulated. Orange roughy represents an ecologically poor seafood choice.

Patagonian toothfish (*Dissostichus eleginoides*), a member of the cod icefish family (Nototheniidae), known in Chile as bacalao de profundidad (cod of the deep), is another deepwater species that lives in the southern oceans. It is a large, dark gray-black fish, with ferocious-looking teeth, that is caught principally using long lines. Patagonian toothfish live to over 45 years old, and may not reach sexual maturity until 10 or 12 years old. In the last three decades, the species has gone from being an infrequent, accidental catch in Chilean coastal congrio (eel) fisheries that was considered worthless by local fishermen, to being a highly-prized catch that is sought after by pirate fishers in the Southern Ocean.

Each step in the virtual trash-to-treasure transformation of Patagonian toothfish has been chronicled by G. Bruce Knecht, in his fascinating book *Hooked: pirates, poaching, and the perfect fish*.[69] This fish tale began in the late 1970s, when a west coast fish merchant discovered the under-utilized species in a fish market in Chile. He renamed it "Chilean sea bass," a name that he thought would resonate with Americans, and then set about developing a market for it in the US. This mild-flavored, white-fleshed, oily fish possessed many qualities that Americans prefer in a fish. After several years of practically giving away frozen fillets in order to acquaint seafood wholesalers with the species, "Chilean sea bass" became the darling of the seafood industry. It was first used commercially in fish sticks, in place of more expensive halibut. Then it gained acceptance in Chinese cuisine, where it was used as an alternative to more expensive black cod. Its bland taste made it perfectly suited for a wide variety of preparations. Around the time that "Chilean sea bass" fillets started selling on the East Coast of the US, Chilean fishers, who had set their long lines deeper than usual, pulled up a phenomenal catch of Patagonian toothfish, and discovered that toothfish lived in very deep water. By then, technology enabled fishers to locate and catch fish in deep water. As demand for this fish was growing, the supplies of it were growing as well. Fish sellers focused on shipping it fresh rather than frozen, which made it a more expensive product. As restauranteurs became enamored with the versatility of "Chilean sea bass," demand for the fresh fish increased, as did its price.

Over 2 decades, "Chilean sea bass" evolved from being a cheap substitute for high-priced fish to being an expensive, highly-regarded seafood. Displaced fishers from the over-fished northern hemisphere looked to this species from the south for their financial salvation. In 1991 and 1992, before Chile understood the size of their toothfish stocks and before they could develop an intelligent management plan for the species, they allowed foreign fishers to catch toothfish in their waters on an exploratory basis. Without effective monitoring, this essentially allowed fishers an open season on the species. After a peak toothfish catch in 1993, the catch plummeted in 1994. By the time Chile introduced restrictions, their toothfish populations had been decimated, and the fishers moved on to exploit other populations without quotas. By the end of the century, chefs in the US were noticing that fresh "Chilean sea bass" was not always available from seafood merchants, and the fish that were available were about a third the size of fish that they had purchased a few years earlier. This clearly showed that the species had been overfished, and juveniles were being caught and marketed. Currently, there are still a few legitimate, legal fisheries for toothfish. In order to market the fish in the US, documentation certifying that fish were legally caught is required. However, without rigorous inspection of seafood shipments, forged catch documents and/or the fraudulent use of legitimate catch documents both compromise the credibility of any paper trail. Unfortunately, illegal, unreported, and unregulated (IUU) fishing probably accounts for about half the annual trade in this species. Despite national and international attempts to regulate this fishery, pirate IUU catches confound any likelihood of successful management, and are pushing toothfish populations into a serious decline. Because toothfish populations have been severely overfished, and face a very slow recovery, especially while they continue to be illegally fished, Patagonian toothfish, (also known as Chilean sea bass) represents an ecologically poor seafood choice.

Sharks, a diverse class of cartilagenous fishes (Chondrichthyes), which means that their skeletons are built out of cartilage instead of bones, like the bony fishes (Osteichthyes), exemplify another unsustainable fishery. They live in various marine habitats around the world. Like the orange roughy and Patagonian toothfish, sharks grow slowly and live a long time, but compared to bony fish, sharks produce very few young.[70] Although sharks are considered slow-growing, late-maturing fish, within a species, growth rates and age at maturity appear to vary with nutrition, by sex, and with geographic area. Accurate figures on the life span of shark species are not readily available, because aging techniques have not been validated for all species. Whale sharks (*Rhincodon typus*), the largest shark species, are estimated to live 100 years. Great white sharks (*Carcharodon carcharias*) are thought to live up to 40 years. Spiny dogfish (*Squalus acanthias*) have the longest validated life span of sharks, 50 to 75 years,[71] although most dogfish probably don't live more than 25 to 40 years. Many shark species do not reach sexual maturity until 10 to 15 years old. Due to differences in their mating behavior, the reproductive rate of sharks is the lowest of all groups of fishes. Most bony fishes broadcast their eggs and sperm into the water column, by the hundreds or thousands. However, sharks mate belly to belly, and the male delivers sperm into the female, using pelvic fins that are modified into claspers, and fertilization occurs inside the female's body. Many species of sharks give birth to live young, or pups. The length of time the female carries her pups varies with species, ranging from 6 to 22 months. For some species, females bear only one or two pups, every two years. As a result of late maturity and a long reproductive cycle, a female shark may bear a very limited number of pups in her lifetime.

A large number of shark species are fished for their fins, their meat, and other products such as cartilage and shark liver oil. This fishing pressure has imperiled many shark populations. A global assessment of 64 pelagic shark and ray species found that nearly a third (32%) were classified as threatened (either "endangered" or "vulnerable") using criteria established by the International Union for the Conservation of Nature (IUCN) Red List of Threatened Species.[72] An additional 24% of pelagic sharks and rays were designated as "near threatened," and could easily change in status to "vulnerable" when more data become available. One of the most serious fishing pressures on sharks is for shark fins. Traditionally, shark fins were made into a soup that was considered a delicacy. It was consumed only by people of high social status, in southern China. For thousands of years, emperors savored it because it was rare and difficult to prepare. After the revolution, the central government in Beijing looked down on its consumption and what it represented. However, in recent decades, as China has become more capitalistic and personal wealth has increased dramatically, demand for shark fins has grown very rapidly. While previously only the elite could afford to eat shark-fin soup, now it is popularly served, to both impress and honor guests, at events ranging from weddings, to business dinners. The soup is legendary for its texture which is derived from the gelatinous, noodle-like fibers of collagen and elastin in the fin's rays. Shark-fin soup is one of the world's most expensive seafood products. In Hong-Kong, the world center of shark-fin trade, a bowl of high quality shark-fin soup can cost up to $100. The high price encourages continued harvest and trade. Much of the trade is illegal and unregulated. We should commend retired NBA all-star player, 7-foot 6-inch (2.3 m) tall Yao Ming for publicly pledging to stop eating shark-fin soup, and urging others in his native China to do the same, in August 2006. More recently, in 2011, he teamed up with British tycoon Richard Branson to appeal to businessmen in Shanghai to stop the trade in shark fins.[73]

Many shark fins are procured through the finning of sharks, that is cutting the fins off live sharks, and dumping their carcasses into the sea. This is an inhumane procedure, which results in the finned shark sinking to the sea floor and dying. It is wasteful of the shark meat, which poor people in some developing countries rely on for protein, and detrimental to the survival of shark species because of their slow-growing, late-maturing, low-reproductive-rate life style. Shark populations are quick to decline, and slow to recover. Based on extensive research in the Hong Kong shark-fin trade, scientists have recently estimated the number of sharks finned annually at 26 to 73 million.[74] The blue shark (*Prionace glauca*) was identified as the dominant fin in the market, followed by shortfin mako fins (*Isurus oxyrinchus*), silky shark fins (*Carcharhinus falciformis*), sandbar shark fins (*Carcharhinus obscurus*), bull shark fins (*Carcharhinus leucas*), hammerhead fins (*Sphyrna* spp.), and thresher shark fins (*Alopias* spp.). The decimation of shark populations in order to satisfy a human desire to impress wedding guests and business associates is particularly sad considering that sharks are one of the oldest groups of vertebrates on earth, and have occurred in our oceans for 350 to 400 million years. Ecologically speaking, shark-fin soup is clearly a very poor seafood choice.

Another fishery that doesn't make sense ecologically is the fishery for seahorses. Although seahorses are too small to be a significant food item for humans, tons of seahorses are harvested annually for use in Asian traditional medicine, particularly traditional Chinese medicine (TCM).[75] The Convention on International Trade in Endangered Species (CITES) estimates worldwide trade at 20 million individual fish. However, Kealan Doyle, an Irish marine biologist posing as a TCM supplier, secretly filmed seahorse-trade activities at clinics, health stores and wholesalers in southern China, for a documentary. As a result of this project, Doyle estimates that TCM annually uses at least 150 million fish.[76] In a chapter of *Poseidon's Steed: The story of seahorses, from myth to reality*[77] entitled "A seahorse cure," Helen Scales tells the eye-opening story of the use of seahorses in TCM, in the past and continuing to the present. Asians have looked to the healing power of seahorses for hundreds of years, at least since the Ming Dynasty in 14th to 16th century China. As the Chinese population of 1.3 billion accounts for about 20 percent of the world's population, we can't lightly dismiss their traditional medicinal practices. Additionally, because outside China, TCM is expanding as an alternative to, or complement to Western medicine, we shouldn't marginalize its global impact.

To the ancient Chinese, everything could be explained by the balance between Yin and Yang energies. Yin forces, the dark side of the Yin-Yang symbol, represent cold, contracting, downward, passive, and weak energy flow. On the other hand, Yang forces, the light side of the symbol, represent active, bright, expanding, hot, strong, and upward energy flow.

In TCM, seahorses are considered a powerful source of Yang energy. Seahorses are consumed as a tonic food to improve the sweet and warm, Yang energy flow. Generally, they are dried, then ground into a powder, which can be put into various medicinal preparations. Although their principal medicinal uses vary from region to region, seahorses are widely used to treat impotence. They are used as aphrodisiacs, and to treat ailments ranging from asthma to heart disease, insomnia, and incontinence. TCM practitioners also use seahorses to facilitate childbirth, to treat thyroid dysfunction, skin problems, and kidney ailments. Unfortunately, double-blind clinical studies that could determine whether seahorses provide measurable medical benefits have not been conducted, and the active ingredients in seahorses which may provide these benefits have not been identified. Once any active compounds are identified, they could be synthesized in a lab, which could reduce the worldwide demand for more seahorses.

Although there are a few, small, dedicated fisheries for seahorses, such as the lantern fishery of Handumon, Jandayan Island, the Philippines, many of the seahorses destined for TCM are caught incidentally in the course of another fishery. For example, when they are picked out of shrimp trawl nets that have been dragged near the sea floor. Populations of seahorse species are hard to quantify, because of their habits, habitats, and cryptic appearance. However, as of 2012, Project Seahorse (IUCN's authority for the seahorse family, Sygnathidae) reports 38 seahorse species were listed on the IUCN Red List of Threatened Species. Populations of 29 species were "data deficient" to further define their status, one species was defined as "endangered," seven seahorse species were classified as "vulnerable," and one species was listed as "least concern."[78] Fishers have recognized a decline in both the size and number of seahorses caught over the last few decades. From reduced landings of seahorses, and undiminished demand for them, it is clear that many populations are shrinking. Choosing any TCM remedy that contains seahorse is ecologically insupportable.

Drastically declining populations and continued high demand make bluefin tuna fisheries unsustainable around the world. Bluefin tuna have evolved remarkable adaptations to suit their life in the open ocean, including sleek, hydrodynamic lines, retractable fins, giant size, strength, speed, and warm-blooded circulation. However, these adaptations are no match for the world's nearly insatiable appetite for their delectable, flesh. All three bluefin species, Atlantic bluefin tuna (*Thunnus thynnus*), Pacific bluefin tuna (*Thunnus orientalis*), and Southern bluefin tuna (*Thunnus maccoyii*), are relatively slow growing, late maturing, and long-lived and thus vulnerable to overfishing.

Although Atlantic bluefin tuna have been fished in the Mediterranean Sea for thousands of years by a variety of cultures, including ancient Greeks, Romans, and Phoenicians, in the western Atlantic, as recently as the 1960s, bluefin meat was considered only suitable for pet food.[79] Without adequate refrigeration, the red meat was prone to deteriorating quickly. Giant bluefin tuna caught off New England in the 1960s were sold for a pittance, five cents a pound, destined largely for cat food. Changing Japanese palates following World War II, and significant advances in refrigeration together resulted in a new, sophisticated appreciation for the taste of raw bluefin flesh. By the 1970s, cargo planes began flying bluefin giants that were caught in the western Atlantic to Tokyo from New York and Boston, paying fisherman significantly higher prices. In short order, fish that had previously been worth $40 to $50 began earning fishermen $1000, when destined for sushi in Tokyo. During the 1980s, at a time when more Americans were embracing sushi as part of a healthy diet, off-the-boat prices of more than $10,000 for a giant bluefin were not uncommon. For many fishermen, catching a giant bluefin was much like winning the lottery. All recent record prices for bluefin have been paid at the New Year's tuna auctions at the Tsukiji fish market in Tokyo. In January 2012, a record price of $736,000 was paid for a 593-pound (269-kg) bluefin.[80] This comes to $1238 per pound. It should be noted that while Japanese diners love bluefin tuna, this special auction price was for a giant bluefin caught near the tsunami-ravaged Japanese shoreline, and reflected a bidding war for publicity, not business as usual. The previous record was $396,000 for a 754-pound (342-kg) bluefin at the New Year's auction in 2011.[81] Fishermen aren't paid prices like this. Rather, in 2011, the average (general category) off-the-boat price American bluefin tuna fishermen were receiving was merely $9 per pound.[82]

Over the last few decades, while breeding stocks and catches of Atlantic bluefin tuna have declined, fishermen and environmentalists have pointed fingers at each other, and pushed hard to influence the agency responsible for managing the fish, the International Commission for the Conservation of Atlantic Tuna (ICCAT). Although politics and economics have frequently trumped science in management decisions, dedicated environmentalists like Carl Safina, have worked to encourage ICCAT to support listing the Atlantic tuna as an endangered species with CITES.[83] Such a listing would stop international trade of bluefin tuna. But Japan, the world's largest consumer of bluefin tuna, has continued to successfully block this designation.

In April 2010, the blowout of the Deepwater Horizon oil well in the Gulf of Mexico presented an additional obstacle to the recovery of Atlantic bluefin tuna stocks. The blowout occurred on the spawning grounds of western Atlantic bluefins, during their spawning season.[84] Surveys of bluefin tuna larvae may provide the first clues as to the impact of the blowout on year-class strength, but the full impact on the fishery won't be realized for years.

At present, all three species of bluefin are listed on the IUCN Red List of Threatened species.[85] Atlantic bluefin tuna has been listed as "endangered." The status of Southern bluefin tuna has been classified as "critically endangered," and Pacific bluefin tuna has been listed as "least concern." Clearly, bluefin tuna cannot be considered a sustainable seafood choice.

Overfishing and/or the collapse of fisheries can result from several factors. Some fish stocks have life histories that render them inherently vulnerable, biologically. Environmental degradation can also contribute to the collapse of some fish stocks. Inadequate knowledge about the population size and age structure of fish stocks, and their mortality and recruitment rates, in combination with inaccurate scientific assumptions can result in poor management policies.[86] As managers attempt to assuage politicians and fishermen, even the best scientific data can be morphed into fishery management decisions that are not ecologically sustainable.

Economic issues tend to complicate questions about sustainable fisheries. As we have mentioned before, too many boats are fishing. This abundance of boats is called fishing overcapacity, excess capacity, or overcapitalization. Overcapacity is an economic issue that reflects inefficient use of resources. If all boats in an overcapitalized fleet were to fish at their maximum fishing effort throughout a season they would exceed the desirable level of harvest or total allowable catch.[87] Many countries contribute to this problem through large financial subsidies to the fishing industry, because governments view fisheries as much more than a food source. Fisheries represent an important source of international trade revenue and a vital source of employment. At present, around the world, tax revenues from outside the fishing industry are being used to subsidize fisheries. Government fishing subsidies can take many forms including direct assistance to people who are under-employed in the fishing industry through unemployment insurance, supplemental-income payments, and retraining; low-interest or guaranteed loans to facilitate the purchase of fishing boats and equipment, or to modernize vessels; tax breaks, including exemptions from paying fuel taxes; investments in infrastructure, including access fees, and port and harbor facilities; economic support through marketing, promotion, and minimum-price guarantees; as well as fisheries management, stock enhancement and conservation programs. Although some government fishing subsidies contribute to sustainable fishing practices, others clearly do not. Global fishing subsidies for 2003 were recently estimated to be between 25 and 29 billion dollars.[88] Worldwide, governments are paying billions of dollars to the fishing industry to continue fishing when it is otherwise unprofitable. Subsidies that support expansions of fishing fleets and overcapacity contribute to non-sustainable overfishing. Even fleet reduction projects that appear to remove excess capacity don't always succeed. Although a particular boat that is bought out of a fishery may no longer fish in that fishery, the "retired" boat is often transferred to another country, or switched to fishing for another species, an "under-utilized" species. Developed countries often pay fees to poor, developing countries for access to their EEZ waters, as distant-water fishing fleets deplete previous fishing grounds. These access fees enable the fishing fleets from developed countries to deplete the fish stocks of developing nations, for example along the coast of west Africa.

Overcapacity and illegal, unreported, unregulated (IUU) fishing are two great problems confronting fishery managers. The management of a fishery is difficult enough when all the participants respect regulations. However, any improvements in the status of managed fisheries can be lost through IUU fishing, when applicable laws, regulations, and standards are disregarded.[89] Illegal activities include fishing out of season, fishing in closed areas, fishing for species that are protected, exceeding quotas, fishing without permission, and disregarding all regulations. Essentially, IUU fishing is pirate fishing, and catches are generally sold on the black market. Some IUU fishing may be directly related to overcapacity, and may be done by dispossessed fishers in an attempt to make a living following fleet reductions in their traditional or national fisheries. Together overfishing, overcapacity, and IUU fishing result in inefficient and wasteful use of resources, and dangerous reductions in fish populations.

What is bycatch and why is it important?

Bycatch is a term used to describe the unintentional (or incidental) catch of something other than the principal target of a fishery. Bycatch results when fishing gear is not selective. It includes the capture of non-target species as well under-sized individuals of the target species. Non-target species that are caught as bycatch in various fisheries include ecologically and commercially important fish species, endangered species, low-value species, and air-breathing species that we empathize with, such as sea turtles, and dolphins. Much of the bycatch is discarded at sea, and not even recorded. Reasons for dumping bycatch range from catching the wrong species, size, or sex (including prohibitions related to season, or gear), to lack of storage space, high grading, and exceeding quotas.[90] To emphasize the destructive nature of bycatch, it is frequently referred to as bykill. Unfortunately, juvenile stages of commercially important fish species often comprise a large portion of the discarded bycatch. Heavy bycatch mortality of juveniles can compromise future generations of those species. A diverse array of marine species are victims of bycatch each year, including thousands of tons of fish and thousands of marine mammals, seabirds, and sea turtles.

Despite attempts and/or regulations to release the non-target species that are landed on a fishing boat alive, they are often ecologically dead, and doomed to die or fall victim to their predators before they regain their physiological and behavioral balance in the sea, as a result of thermal shock or barotrauma (a pressure stress that causes physical damage when the air bladder of a fish expands rapidly as they are pulled to the surface from high-pressure environments deep under water). In that sense, vigilance, and rapidly returning non-target species to the sea doesn't necessarily insure their survival or reduce the bykill. However, gear modifications (bycatch reduction devices) that enable non-target species to escape from trawl nets while the net is being towed, or before it is hauled up, often result in the survival of the non-target species that exit the gear through escape panels, thus reducing bycatch.

The capture of dolphins by the Pacific tuna fishery is probably the most widely known example of commercial fishing bycatch. Early on, tuna were caught individually, with hook and line, without harming dolphins. However, tuna fisheries switched to using purse seines, and this resulted in the bycatch mortality of large numbers of dolphins. For reasons unknown to us, large yellowfin tuna (*Thunnus albacares*) often swim beneath pods of dolphins. This ecological association of dolphins with schools of yellowfin tuna has resulted in the unintended mortality of millions of dolphins. Purse-seining fishing boats utilize spotter airplanes to look for dolphin pods in order to locate schools of tuna. A small boat tows the end of the purse seine, encircling the tuna and dolphins with a huge net, which is towed back to the main fishing vessel. From there, the seine is pulled in using sophisticated hydraulics that cinch up the bottom and sides of the net, and haul in everything that has been encircled by the net. Unless fishermen are especially vigilant, dolphin bycatch mortality results when they become entangled in the nets and drown. Public outrage against the needless destruction of dolphins has led to a significant reduction in bycatch losses. Estimated dolphin mortalities through tuna fishing decreased from 133,000 in 1986 to less than 2,000 in 1998.[91] Increased vigilance by fishermen and improved fishing techniques now enable trapped dolphins to escape from nets. This represents a bycatch-reduction success story. In response to consumer and environmentalist demands, cans of tuna sold in the UK and US now bear "dolphin-safe" labels, reflecting the adoption of dolphin-friendly fishing practices. However, some international fisheries continue to endanger dolphins. And longline fisheries continue to result in dolphin bycatch mortalilties. Unfortunately, some of the measures that have been adopted to reduce dolphin bycatch have unintentionally switched bycatch to other species such as sharks, and sea turtles.

Bycatch mortality from pelagic longline fisheries represents the most serious global threat to pelagic seabirds, especially albatrosses and petrels. Longline fishing (or longlining) is a major fish-capturing technique used by commercial fishermen that is named for the gear that they deploy: a main line that may be several miles/kilometers long, that carries as many as several thousand shorter, branch lines with baited hooks. The baited lines are fished from the stern of fishing vessels. Longline fisheries catch seabirds that try to seize bait from hooks before, or soon after, the hooks drop into the water. Seabirds become hooked or entangled in the lines while the gear is being set. They get dragged under as gear sinks, and drown. Worldwide, longline fisheries have resulted in the bycatch mortality of more than 60 seabird species, including several endangered albatross and petrel species.[92] Recent estimates of observed, reported bycatch place the total number killed annually in longline fisheries at 160,000 to 320,000 seabirds,[93] but the actual number could be much higher as a result of illegal, unreported fishing. In the Alaska demersal longline fishery, bycatch mortality of the short-tailed albatross (*Phoebastria albatrus*), an endangered species, is considered so serious that the fishery faces closure when the bycatch of short-tailed albatross exceeds 6 over a two-year period.[94] Not surprisingly, seabird bycatch mitigation is required on vessels in this fishery.

Many options exist to reduce seabird bycatch from longline fishing.[95] Longliners can modify their gear, bait, and the timing and location of fishing. Gear modifications include using bird-scaring lines, increasing the sinking rate of baited lines with heavier line weights, using chutes to set baited lines below the water line, using modified hooks, and increasing the distance between hooks. Bait modifications include using only thawed bait which sinks faster than frozen bait, piercing the swim bladders of bait to improve sinking rate, and dyeing bait blue to reduce its visibility to birds. Temporal/spatial modifications of fishing include fishing at night, while ship lighting is kept low to avoid attracting birds, and seasonal or regional closures of fisheries to avoid longline fishing when colonial seabirds are foraging to feed nestlings. The strategic discharge of fish offal is another way to reduce seabird bycatch. Seabirds feed on discarded offal. If fish wastes are either dumped away from the stern where the lines are set, while lines are being set, or if they are not dumped overboard at all, seabirds mortalities can be reduced.

Perhaps the most successful deterrent for seabird bycatch used in longline fisheries is bird-scaring streamer lines or tori lines. Tori means bird in Japanese. Tori lines were developed by Japanese bluefin tuna fishermen to frighten seabirds away from bait while longlining for tuna. Their motive was to reduce their loss of bait, and hopefully increase their tuna catches. The tori lines helped their bottom line while decreasing seabird mortality. Pairs of tori lines are strung between poles on the stern of a vessel and floating buoys that are towed behind the vessel. Colored streamers are attached to the lines. They flap erratically in the wind, above the site where baited longlines enter the water. When tori lines are used in conjunction with properly weighted longlines, baits sink quickly and seabird mortality is virtually eliminated. By the time the bait has passed the bird-scaring streamers, it is deeper than the diving range of most seabirds.

Besides seabirds, longline fisheries for tuna and swordfish have another bycatch problem: sea turtles that get hooked after eating a bait, and then drown. However, progress is being made on making these fisheries more turtle friendly. For example, bycatch of turtles can be minimized if the lines are fished deeper than turtles normally forage. Switching from the traditional J-shaped hooks to a new large, nearly circle-shaped hook has been shown to significantly reduce bycatch of loggerhead sea turtles.[96]

Clearly longline bycatch mortality can be reduced if fishermen modify their techniques, and many countries have adopted longline fishing techniques that reduce bycatch. Unfortunately, the oceans abound with illegal, unreported longline fishing that is conducted outside the reach of any regulations.

In the shrimp-trawl fishery, the biomass of discarded bycatch is generally many times larger than the biomass of the target species. A 2005 analysis of the shrimp fisheries in the Gulf of Mexico reported that discarded biomass weighed 4.6 times more than shrimp landings,[97] but this represented an improvement over the previous decade. One truly unfortunate aspect of bycatch is that juvenile stages of commercially important fish species are often caught and discarded as bycatch in another fishery. In the Gulf of Mexico, the shrimp-trawl bycatch of juvenile red snapper (*Lutjanus campechanus*) represents a large source of mortality in red snapper populations. Just how large? Between 1992 and 1996, the average number of juvenile red snapper taken as shrimp-trawl bycatch in the Gulf of Mexico, was estimated at 26 to 32 million fish per year.[98] Bycatch reduction devices that are mounted on shrimp-trawl nets allow non-target marine species to swim out and escape from the net before it is pulled in to the boat. On trawl nets, bycatch reduction devices (such as "doors" or "windows") that take advantage of behavioral differences between the target species and non-target species are among the most successful at allowing non-target species to escape.

Shrimp fisheries also have problems with bycatch of sea turtles. Because all sea turtle species are considered either threatened or endangered, fisheries need to minimize their impact on sea turtle populations. Turtles can get caught in shrimp trawls and drown. Although turtle excluding devices (TEDs) are mandatory on nets of US trawlers and foreign vessels selling shrimp in US markets, they are not always used. Basically, a TED consists of a grid of bars that permits easy passage into the tapered, collecting end (the codend) of the trawl net for shrimp, but directs larger animals, such as turtles, through a trap door in the back of the net that allows for their escape.[99]

One other significant bycatch problem has been the loss of marine life caused by abandoned gear continuing to fish, or ghost fishing. Gear can be lost, abandoned (intentionally or not), or otherwise discarded at sea for a variety of reasons, including enforcement pressure, economic pressure, spatial pressure, and adverse environmental conditions.[100] Lost longlines and trawl nets result in the loss of their initial catch, but longlines cease fishing as soon as their bait is gone, and trawls eventually sink to the bottom and cease fishing. However, some types of lost gear such as gill nets can continue to fish passively (ghost fish), catching and killing various marine species, including dolphins. Ghost-fishing gill nets may move through the water column, sinking when they are weighted down by their catch, then rising as the catch decomposes or is eaten. Fish that die in the net may act as bait and attract more fish. Historically, nets were made of materials that would eventually decompose, and stop fishing, however modern gear is constructed of synthetic, non-biodegradable fibers. A lost gill net may continue to fish for up to ten years. Lost fish traps that are constructed of synthetic materials can also continue to fish for years. Fish that die inside ghost-fishing traps act as bait, and attract more fish. To alleviate prolonged ghost fishing by lost traps and pots, new traps are being constructed with biodegradable or corrodible ties that secure escape panels, or with galvanic timed release escape panels.[101] Once the ties or panels break down, fish can escape, and the trap will cease to ghost fish. Most trap and pot fisheries around the world now require some type of ghost-fishing reduction device.

Bycatch poses significant problems for fisheries around the world. Many countries have attempted to reduce bycatch through gear modifications, collectively named bycatch-reduction devices, and through modifying the behavior of fishers. However, few countries have been as successful and direct as Norway in addressing bycatch. Norway has a deep-rooted fishing tradition, and to some extent, fishing is a part of their Viking-spirited, national identity. They take fishery resources seriously, and are in the forefront of fisheries science and management. Norway has outlawed the dumping of bycatch at sea, and requires the landing of an entire catch.[102] No fish are discarded at sea. This policy encourages selective fishing, but also makes use of essentially everything that is caught. Fishermen are compensated for a portion of the value of unintentionally-illegally-caught fish, forfeiting the remainder of the commercial value to the state. Thus by landing bycatch, fishers can recover costs that would have been lost had they dumped the unintentional catch at sea. In Norway, bycatch species have been creatively marketed, and virtually everything is used and considered of value.[103] However, fishers earn the most money when they catch their target species. Thus, by mandating the landing of bycatch, Norway has at the same time encouraged their scientists and fishers to develop more selective gear, which reduces the bycatch of non-target species.

What are the ecological impacts of capture fisheries?

Beyond collateral mortality through discarding the bycatch of undesirable species and year classes, through high grading of the target species (i.e., keeping the largest fish while dumping smaller, but legal-sized fish), and through ghost fishing, capture fishing has too frequently resulted in overfishing of the target species. Basically, overfishing is removing more fish from a population than can be replaced by young fish as they grow and recruit into the fishery. Overfishing happens for a variety of reasons including too many fishers in open access zones, poor management, government subsidies to the fishing industry, advances in fishing gear, and technological developments in both fish finding capabilities and global positioning systems that leave fish with virtually no place to hide, advances in at-sea processing, pirate fishing, and greed. High prices and high market demand fuel overfishing of species such as bluefin tuna. Regardless of why overfishing occurs, the removal of vast numbers of fish can shift both the structure and function of marine ecosystems. Capture fishing can significantly alter marine food chains through the virtual removal of large populations of prey or predator species. If small pelagic fishes are overfished, then all the species that are linked to this species via the food web will be impacted. The collapse of prey fish populations can result in increased mortality and population declines of higher trophic levels, including top predators. On balance, the loss of a trophic level within an ecosystem triggers cascading trophic effects that may shift the community structure to a new regime.[104] During years when forage fish populations fail to provide enough food, nesting marine birds are forced to abandon their nests and try to survive to breed another year. Populations of marine mammals and predatory fish also decline in response to the collapse of forage fish populations.

Capture fishing utilizes a wide range of fishing gear and/or fishing techniques that degrade marine habitats. These have been collectively named destructive fishing practices. Any type of mobile fishing gear that is dragged across the bottom damages the sea floor. Commercial bottom trawls that are used for shrimp and groundfish and dredges that are used primarily to dig up scallops and clams are the notorious "thugs" of fishing gear. Unmodified, they both leave trails of destruction across sensitive sea floor habitats. Comparing the effects of various bottom-fishing techniques: intertidal dredging is more damaging to benthic habitats than scallop dredging, which is more damaging than intertidal raking, which is more damaging than trawling.[105] Although this comparison minimizes the damage caused by trawl nets relative to heavy metal gear that grabs huge bites out of sea floor habitats, bottom trawling is far from benign. Historically, bottom trawling was used only on relatively smooth bottom habitats. The weight of the gear (the trawl net, its cables, and the large doors that keep the net open) further smooths out irregularities in the sea floor as the gear is dragged forward by a fishing vessel. Repeated trawling changes sea floor communities, breaks down both physical and biological structural components of habitats, and reduces the productivity of bottom habitats. Until quite recently, trawlers avoided fishing on structurally complex sea floor habitats, including boulder fields, jagged rocky pinnacles, and coral reefs, because they didn't want to tear their nets, or lose their gear. However, in the 1980s, following the introduction of heavy rollers and rock hoppers to the foot ropes of trawl nets (the base lines at the mouth of the net openings, that hold the nets to the sea floor), fishers returned to the grounds that they had avoided, driving more powerful boats. The rollers and rock hoppers bounced over, rolled over, or displaced massive obstacles, and crushed deep sea corals, dragging heavy trawl nets behind, along the sea floor. Adding these accessories to their trawl gear enabled fishers to access previously inaccessible, structurally complex areas such as the pinnacles of underwater seamounts without damaging their nets.[106] In the short term, their catches increased, but in the long term, catches decreased due to a loss of sanctuary for bottom-dwelling marine species.

Bottom-fishing gear can reduce habitat complexity and perturb benthic sea floor communities. These gear plow across the seabed, leaving furrowed tracks behind. Towed and dragged bottom-fishing gear stir up bottom sediments, thereby altering the grain size of remaining sediments and increasing turbidity in the water column. Beyond mere physical disturbance, these changes can seriously affect nutrient cycling and primary production. Sea floor habitat complexity is a function of the surface topography on the sea floor, internal structure, and large sessile, structural organisms or colonies (like coral or sponges). High habitat complexity provides lots of refuge space, or safe zones, for a variety of prey and juveniles of many species, and leads to stable communities with high biodiversity. Bottom-fishing gear can damage benthic habitats so substantially that the fish stocks that depend on them cannot recover. Physically, this damage ranges from the disturbance and re-suspension of sea floor sediments to the complete destruction of physical relief, via flattening habitat structure or through the removal of substrate and structural fauna. Habitats with a high proportion of structural fauna are likely to be the most sensitive to the physical disturbances that result from active bottom fishing using towed or dragged gear. Vulnerable benthic habitats include coral reefs, maerl (calcified red algae) beds, mussel and oyster beds, sponge beds, sea fan aggregations, and sea grass beds.[107] Repeated gross perturbations of the sea floor, such as inflicted by destructive fishing practices, can alter both the form and function of the seabed, and transform vigorous communities into virtual dead zones. To put this in perspective, bulldozing down an old-growth forest can be considered a terrestrial analogy to the removal of relief and structural organisms from deep sea habitats by dragged bottom-fishing gear.

Even if the gear and techniques used by commercial fishers are responsible for most of the ecological damage caused by fishing, in shallow-water habitats, some artisanal fishers have also left a trail of ecological damage. The techniques used by any individual artisanal fisher may have a small effect, however due to the sheer number of artisanal fishers, collectively they can have large effects on benthic habitats. Tropical coral reef ecosystems have the greatest habitat complexity and richest biodiversity of any marine environment. Unfortunately, species-rich coral reef ecosystems also represent the primary fishing grounds for tens of thousands of artisanal fishers. Although some native fishermen continue to the use the sound traditional fishing methods that have been practiced for centuries by their ancestors who had remarkable knowledge of marine ecology and fishing skills,[108] quite a few other artisanal fishing methods are destructive.[109] For example, throughout the tropics, artisanal fishers damage coral reef habitats when they use drive netting techniques. One method, which originated in Japan, known as muro-ami, has been used throughout southeast Asia and the Philippines. Muro-ami involves a number of fishers who generally swim above coral reefs while repeatedly, rhythmically dropping heavily-weighted, scare lines on the reef to drive the fish out of the reef and into their encircling net.[110] The weights may be cement blocks or rocks. On deep reefs, swimmers drop evenly-spaced, scare lines that are decked out with lateral streamers to create a visual wall. Sometimes, fishers bang metal poles on the reef, driving reef-associated fishes to abandon protective reef structures and flee into the bag end of their net. In very shallow water, fishers often break off coral branches to drive out fish that are closely associated with branching corals. Coral heads that are repeatedly pounded by weighted lines and metal poles can easily be reduced to rubble. Although, muro-ami fishing removes a large biomass of fish from coral reef habitats, many of the lost fish could be replaced fairly quickly if the ecosystem was healthy, and new individuals recruited into the habitat. However, rebuilding coral reef structures that have been reduced to rubble takes a long time. Muro-ami fishing has been banned in some areas, due principally to the endangerment of children who are employed as divers and swimmers. Legal or not, the practice continues.

Another technique used by artisanal fishers that wreaks havoc on reef habitats is blast fishing, i.e., fishing using explosives. Worldwide, blast fishing is a great and immediate threat to coral reef ecosystems. Fishing with explosives has occurred for over a century, especially off China, a country notable for its role in the development of explosives. In the Indo-Pacific region, which is considered the epicenter of coral reef biodiversity, blast fishing took off after the end of World War II. Although blast fishing has been banned in most countries, the practice continues, to some extent, on southeast Asian, east African, and Indo-Pacific coral reefs.[111] Explosives are fairly easy to obtain. They can also be assembled using cheap, readily available raw materials, such as fertilizer, kerosene, and a simple fuse. Unexploded ordnance from past and current conflicts has provided a free source of explosives for some fishers. In southeast Asia, as fishery resources have declined, due to overfishing, pollution, and dredging, desperate, impoverished fishers have taken to fishing with explosives. They risk their lives hoping to catch a lot of fish in a short time. Blast fishing typically targets schooling reef fishes and small pelagic species. Once a school is sighted, a lighted bomb is thrown into the school. After the bomb explodes, fish are killed or stunned by the shock wave. Dead fish that float to the surface are scooped up, but other fish sink when killed. Free-diving or hookah-rigged divers enter the water after a blast and gather up their catch, but a bykill of non-target fish and invertebrates results. Fish caught via blast fishing may end up quite damaged in appearance, depending on how close they were to the center of the explosion. Unfortunately, in China, it appears that some consumers may actually prefer blast-damaged fish, because they are considered safer to eat than fish caught in heavily-polluted inland waters. Blast fishing causes substantial damage to the physical structure of the reef, creating a crater and often reducing a reef to rubble. It also causes extensive damage to the biological components of reef communities. Blasted reefs support a lower biomass of fish, and fewer fish species, which translates into less fishery production. Extensively bombed coral reefs have virtually no way to recover from repeated blast fishing. Chronic blasting can result in shifts in ocean currents, creating unstable habitats that are hostile to new coral recruits. Bits of coral rubble, moving freely in response to ocean currents, abrade and smother coral larvae that attempt to settle.[112]

Artisanal fishers also fish by poisoning coral reefs. In some areas, fishing using poisons is a traditional method. However, as fish have become more scarce, fishers have moved away from their traditional, plant-derived poisons, and toward using inexpensive sodium cyanide or pesticides. In the Indo-Pacific region, cyanide fishing began as a method to collect fish for the aquarium trade in the 1960s, but by the 1980s, fishers realized that a great deal money could be made supplying large, live, reef fish to restaurants in Hong Kong, Singapore, and China. Cyanide fishing for the live-seafood restaurant trade principally targets grouper species (in the family Serranidae), Napoleon wrasse (*Cheilinus undulatus*, also known as humphead wrasse), a threatened species[113] in the wrasse family (Labridae), and rock lobsters (*Panuliris* spp.).[114] Diving fishers squirt cyanide into the water around a coral reef structure to stun reef fish, which are then easier to catch without being visibly damaged. Fishing with cyanide continues, even in countries where it has been outlawed. Beyond stunning the target species, poisoning can kill coral polyps, algae, and various biological components of reef communities.

In any habitat, flora and fauna have the ability to adapt to some level of physical disturbance. Marine communities differ in their resilience to the physical and ecological disturbances caused by bottom fishing. Shallow-water communities that are adapted to frequent physical disturbances, such as hurricanes, are more resilient to the effects of bottom fishing than stable, deep-sea habitats. The ecological effect of any disturbance by bottom-fishing gear is related to the frequency and magnitude of natural disturbances that occur in a habitat. Marine communities that have evolved with a very low level of physical disturbances in virtually unchanging, stable habitats, such as the deep sea, have little resilience to the damage caused by bottom-fishing gear. Both the longevity and severity of ecological effects of fishing are magnified in stable habitats. Intense bottom fishing can result in substantial reductions in both the size and number of fish and invertebrates in marine habitats. Many other types of fishing gear can wreak marine ecological havoc as well, including gear that is snagged and lost on bottom features, such as deep-sea coral and sponges; and ghost-fishing gear that was constructed from non-biodegradable materials.

What about aquaculture?

Aquaculture is an ancient practice of farming aquatic plant and animal species. It encompasses marine aquaculture, also known as mariculture, the farming of various marine organisms, including fish, shellfish, and seaweed, in seawater or brackish water (tidal water with a salinity that fluctuates between seawater and freshwater). Early aquaculture began in China, about 4,000 years ago.[115] Through the millennia, aquaculture has expanded in both extent and resource expenditures. Traditional aquaculture typically involves simple, inexpensive, low-maintenance, minimally-supplemented, "extensive" practices, through which low stocking densities result in low yields, in semi-natural environments. Planting oysters to grow out on the bottom of a bay, or off-bottom in bags or nets suspended from racks in a bay, are examples of simple, low-tech, extensive aquaculture. The oysters feed themselves by filtering edible particles out of the water, and grow un-aided until harvest.

In contrast, modern, high-tech aquaculture typically involves complex, resource-costly, high-maintenance, highly-supplemented, "intensive" practices, through which high stocking densities result in high yields. In an over simplified sense, fish and shellfish in intensive aquaculture live in a complete life-support system. They grow to market size in a crowded, intensive-care environment that provides all their nutritional requirements. To obtain good yields, promote growth, reduce stress, and control disease, very good water quality must be maintained. Growing Atlantic salmon in floating net pens in Norwegian fjords is an example of intensive aquaculture. In response to rising demands for quality seafood products and aided by technological advances, modern aquaculture has grown rapidly in recent decades. While at first it may sound like a great idea, a closer examination is necessary before we jump to any unreasonable conclusions.

Worldwide, seafood production comes from one of two sources. It is either harvested from an aquaculture facility or captured in the wild. A comparison of aquaculture with capture fisheries reveals that both sources of seafood have positive and negative aspects. We need to weigh them against each other to figure out which product is ecologically the healthier choice, both locally and globally.

A brief comparison of farm-raised fish with wild-caught fish illustrates the tradeoffs between cultured and wild-caught seafood. Perhaps the biggest advantage of farmed fish is that aquaculture can supply a uniform product on a reliable schedule. The size at which fish are harvested is determined by what markets demand. Mass production results in predictable yields, at relatively low cost, often year-round. Farmed fish tend to have a higher fat content than wild fish, which is considered desirable in Japanese markets, however the chemical composition of the fat in farmed fish may not be as healthy as the fat of wild fish. One drawback of farmed fish is that high-density, intensive fish culture stresses the fish, and makes them vulnerable to parasites and diseases, both of which often require treatments with harsh chemicals. These treatments, excess nutrients, and fish wastes result in localized pollution of the marine environment. Although the pollution generated by fish culture may be confined to the proximity of culture facilities, aquaculture's impact on marine ecosystems stretches globally, especially with regard to the vast volume of high-protein fish meal used in the specialized-formula feed that farm-raised fish consume.

Wild fish have the advantage over farmed fish of being a natural, organic product. And organic foods are both good for us and good for the environment. Wild fish that are caught in the most pristine seas, are a superior quality product. They tend to be lower in fat, and free from the chemical residues that farmed fish might carry. Wild-caught fish taste better, and have a firmer texture. In the case of salmon, wild fish achieve their beautiful color naturally, through the food chain, not through the ingestion of food coloring additives like farmed salmon. One major disadvantage of wild-caught fish is the unpredictable nature of capture fisheries. The supply of wild fish can be both temporally and spatially sporadic, due to a number of factors including season, weather conditions, population fluctuations, and regulations. Also, wild fish vary tremendously in size, and are not always exactly the size buyers are looking for. Wild fish are much more expensive than farm-raised fish. Even though the high prices of wild-caught fish reflect the fisher's expenses of operating a fishing vessel, the true cost of capture fisheries is often hidden under substantial governmental subsidies. Subsidy intensity varies with geographic region, with recent estimates ranging from about 30 percent of the landed value of the catch in Asia and North America to over 50 percent in Africa.[116]

Although aquaculture can produce some pretty amazing yields in terms of biomass produced per acre of pen, at what cost is this yield accomplished? High density, intensive aquaculture expends a tremendous amount of high quality resources. The environmental costs include point-source pollution, the degradation of water quality in the vicinity of facilities, and the introduction of harsh chemicals (antibiotics, fungicides, and disinfectants) used as treatments to combat disease.[117] Residues of antibiotics that are fed to farmed fish can be transferred to wild fish, incorporated into benthic communities beneath grow-out pens, and passed on to people who eat the fish. As a result, the use of antibiotics in aquaculture may increase the global risk of developing antibiotic-resistant bacterial pathogens in both fish or shellfish and humans. Some aquaculture techniques may facilitate the transmission of both exotic and native parasites and pathogens to wild populations that utilize adjacent waters. Cultured fish often escape from farmed conditions and they can end up occupying habitats and consuming food resources that are essential to the viability of wild populations. When exotic species are cultured, escapes result in the introduction of new species into an ecosystem. When non-native genetic stocks are cultured, escapes can result in the genetic contamination of local populations.

In floating net pens where salmon are grown out, wastes produced within the pens (fecal material and excess feed) can seriously degrade water quality beneath and down-current of the pens through the production of a tremendous amount of nutrient enrichment, both organic and inorganic. Heavy metals can also accrue in the environs as a result of zinc compounds in feeds, and copper-containing anti-fouling treatments for pens. In the water beneath the pens, dissolved oxygen levels drop as bacteria break down excess feed and fecal material. Nutrient enrichment also reduces water transparency and can cause algal blooms. Blooms can be stinky, and may kill fish.

How does aquaculture impact wild fish populations?

Contrary to public opinion, aquaculture generally does not relieve pressure on ocean fisheries. Instead, wild fish populations are negatively impacted by the practice of aquaculture principally in two extremely important ways: feed and seed.[118] "Feed" defines the ultimate source of the food given to farmed species, in grow-out facilities. Wild fish populations feed aquaculture. Vast schools of small, pelagic, marine fishes are captured in distant seas and then rendered into high-protein feed for carnivorous farmed species. For example, in the case of shrimp, raising 1 kg of shrimp in aquaculture requires about 2 kg of high quality fish meal. By assuming a fish-to-fish-meal conversion rate of 5 to 1, then 10 kg of wild fish would be needed to produce 1 kg of farmed shrimp. In terms of feed, the impact of aquaculture on wild fisheries is very large.

Wild feed

Specially-formulated feeds for cultured species are generally derived from wild populations of small, oily, pelagic fish such as anchovies, jack mackerel, and menhaden that have been rendered into fish meal, fish protein concentrate, or fish oils. Globally, industrial fisheries for these fish land a greater tonnage than white fish fisheries. Peruvian anchovies are the dominant fish caught in the industrial fishery. In the North Sea, sand eels, and capelin are the targets of industrial fisheries. Much of the world's supply of fish meal is produced thousands of miles/kilometers away from aquaculture facilities, in countries such as Peru, Chile, and China. Transporting fish feed vast distances consumes additional resources, and adds to the global costs of aquaculture. The fish-meal industry has environmental costs as well. For example, untreated effluent from fish-meal factories has covered the sea floor of at least one bay in South America (Paracas Bay in Peru) with a thick layer of fat, and spawned a harmful algal bloom that resulted in a sizable dead zone, and closure of the port to all fish landings for 22 days in April 2004. The closure resulted in an estimated economic loss of $27.5 million.[119] In this case, environmental degradation caused by fish-meal factories interfered with local capture fisheries for human markets. When humans consume fish directly, they obtain the maximum nutritional value from fish. Rendering fish that humans could eat into fish meal to feed farm-raised fish uses populations of small pelagic fish species wastefully and inefficiently.

While fish oils that are processed for human consumption are purified to remove contaminants, when fish oils, fish meal, and fish protein concentrate are processed for aquacultural feeds, the same de-contamination procedures are not followed. As a result, farmed fish have higher levels of contamination with PCBs, dioxin, organo-chlorine pesticides and other compounds than wild fish.

One truly frightening aspect of aquaculture is that as it keeps growing, so does its need for high-protein feeds. Presently, the worldwide production of fish meal, fish protein concentrate, and fish oils depends on industrial fisheries for small pelagic fish that may be at or very near the maximum level that can be sustainably harvested on a continuing basis. Increasing aquacultural feed without overfishing may require re-thinking how fish meal is used. Fish meal has been used to supplement diets that are fed to livestock including pigs, cattle, sheep, and poultry – none of which would ever consume fish under normal circumstances, and fed to herbivorous fishes, such as carp and tilapia, that do not need the protein, to boost their growth rates. Although the greatest commercial use of fish meal is for animal feeds, it is also used in agricultural and industrial applications. Fish meal may have become far too valuable as a feed, especially for farmed fish, to continue some of its various and sundry industrial uses. In the case of Peruvian anchoveta (*Engraulis ringens*), fish meal is also incorporated into fertilizer, linoleum, makeup, margarine, and paint.[120] Similarly, in addition to animal feeds, a variety of human food, health, and beauty products are supplemented with fish oils. But the most egregious use of fish oil occurred in Esbjerg, Denmark before fish oil was taxed. Industrial fishers were catching and delivering so many sand lance (*Ammodytes* spp.) to the TripleNine fish protein facility that a glut of fish oil resulted. An enterprising plant manager sold the excess oil to power stations to produce electricity.[121] The fish oil made poor-quality coal burn better. Clearly, whenever fish oil ends up being used as fuel oil, resources have been wasted, and way too many fish have been caught and rendered.

Without a more judicious use of fish meal, increasing fish-meal production to meet the needs of aquaculture in the future, might be achieved in several ways. For example, more fishery wastage could be re-directed into fish meal by using unsold fish or offal (i.e., heads, and carcasses left after fillets are cut off). In 2010, about 36 percent of global fish meal was derived from offal and trimmings.[122] However, trash fish and offal are also used for bait in some fisheries, and diverting these resources into fish-meal production could create a bait shortage. Alternatively, bycatch that would normally be discarded at sea could be processed into fish meal. Both of these ideas seem sound, ecologically, although that doesn't mean either would be simple or inexpensive to implement. One less appealing approach to increasing fish-meal production is targeting non-traditional species for industrial fisheries: using fish that are now commercially harvested principally for human consumption. While a fisheries economist might be able to justify this approach, ecologically it makes little sense to harvest inexpensive fish that poor people in developing countries routinely consume, in order to grind them into fish meal, and then feed the fish meal to more marketable, high-priced, farm-raised fish that will feed affluent consumers.

Another unappealing "solution" to the problem is moving down the food chain and harvesting vast krill populations. Krill have excellent nutritional potential for fish feed. However, these shrimp-like, planktonic, marine invertebrates (in the order Euphausiacea, also known as euphausiids) support a substantial number of marine food webs, particularly in the Southern Ocean, including marine mammals (whales and seals), seabirds (e.g. penguins, petrels, and storm petrels), squid, and fish. Krill numbers appear to be correlated with the extent of sea ice in Antarctica. Sea ice provides a refuge where krill can spend the winter, with ice algae to eat, protected from predators. During periods with reduced sea ice, krill numbers decline. Thus krill populations, and the food chains that they support, may be particularly vulnerable to global warming. Because of the importance of krill in marine food webs in Antarctica, the Convention for the Conservation of Antarctic Marine Living Resources (CCAMLR), was established in 1982, to limit krill harvest.[123] Current krill harvests in the Antarctic Ocean are not threatening marine food webs. However, dramatic increases in commercial krill harvests in conjunction with reduced sea ice could be devastating to the Antarctic marine ecosystem.[124] Increasing either the harvest of krill or fish species that are now consumed by humans to produce high-protein meal for aquaculture could have very sobering global consequences. In order for aquaculture to produce more fish, feeds may need to be carefully reformulated, incorporating more plant proteins when possible.

Wild seed

"Seed" defines the source of the stock that will be grown out, in aquaculture facilities such as pens, ponds, rafts, etc. Often, wild populations are robbed of their future generations as brood stock, fertilized eggs, larvae, or juveniles are captured and diverted into aquaculture. In terms of lost seed, the impact of aquaculture on wild fisheries can be very large. Aggressively harvesting seed stock from the wild can cause the collapse of wild populations. This was seen in the late 1990s when population of European eels (*Anguilla anguilla*) declined following an expansion in the capture of juveniles, at stages known as glass eels or elvers, in order to stock grow-out facilities throughout Europe and in Japan.[125] Because breeding eels in captivity has very rarely been successful, eel culture is dependent on the capture of wild elvers. Wild eel populations are vulnerable partly due to their complex life history that includes a long transatlantic migration of adults. Spawning has never been observed in the wild and wild-spawned eggs haven't been found, but early larval stages have been found in the Sargasso Sea, in the western Atlantic Ocean, south of Bermuda and east of the Bahamas. Spawning is thought to occur somewhere in the Sargasso Sea, and adults are assumed to die after spawning.[126] As the young develop and metamorphose, from fertilized eggs into larvae (leptocephali) and then transparent juvenile stages (glass eels), they are transported by the Gulf Stream back to Europe where they enter coastal waters and river systems. This trip can take up to three years. When they enter fresh water, their skin darkens, and at this stage they are called elvers. At about four inches (10 cm), they enter an immature adult or yellow-eel phase. This phase lasts from five to nine years. European eels can spend 20 years in fresh or brackish water before returning to the Sargasso Sea to spawn. Sexually mature eels are known as silver eels.

European eels are essentially late-maturing fish, that are heavily harvested at several life stages, and all of the stages of eels that are harvested are pre-spawning stages. This basically represents two major strikes against the likelihood that an eel fishery can be a sustainable fishery. In addition, their complex life history makes them vulnerable to anomalous oceanography conditions, as well as to physical, chemical, and/or biological perturbations in coastal and freshwater ecosystems. One further assault on the wild European eel populations has been the harvest of glass eels directly for human consumption in Asian markets. The collapse of Japanese eel (*Anguilla japonica*) populations over the last three decades has fueled high demand for European glass eels in Japan, where they fetch high prices. Unfortunately surging prices have encouraged illegal fishing for European glass eels. Demand for American eels (*Anguilla rostrata*) has increased as populations of European and Japanese eels have declined. To many Americans, eels may sound like an unfamiliar and unappealing fish, however, they are sought after in many cuisines around the world, especially in Asia and Europe. For example, although many people may have only eaten eel as unagi (freshwater) or anago (sea eel) nigiri sushi, eels are considered a traditional Christmas Eve delicacy in Italy; smoked eels are very popular in northern Europe; and jellied eels are sold at street stalls in the east end of London. Due to their complicated life history, and their declining populations, eels are not an ecologically smart seafood choice.

The practice of capturing larval or juvenile stages in the wild, and growing them out in captivity compromises the size of wild year classes or cohorts. Aquaculture requires substantial food resources that depend on primary and secondary production from distant marine ecosystems. In the worst case, juvenile stages of commercially important species are caught in trawl nets, and fed as "trash" fish to larger fish in grow-out pens, or ground up as fish meal and used as feed. I have seen nets full of juvenile fish fed to large grouper that were being held in net pens, in the New Territories of Hong Kong, for the lucrative live fish trade in Hong Kong restaurants.

Partly farmed, partly wild

Although aquaculture and capture fisheries represent distinct sources of seafood production, a gradation exists between these two activities, along which many aspects of fisheries and aquaculture overlap and intertwine. For example, stock enhancement has been undertaken for many wild fishery populations, by releasing young hatchery-reared animals into the wild, to fatten themselves through marine food chains, until they grow into the capture fishery. In general, industrialized countries have engaged in this practice, known as sea (or marine) ranching, with Japan ranching the greatest number of species. Marine ranching has been aimed at rebuilding, maintaining, and increasing the value of fisheries, as well as creating new fisheries for species ranging from flounder in Japan, to red drum in Texas, to sturgeon in the Caspian Sea.[127] Some stock enhancements are considered to be quite successful, such as the enhancement of Texas red drum (*Sciaenops ocellatus*), which was done in conjunction with significant reduction in fishing effort.[128] However, by itself, stock enhancement has been criticized as ineffective at increasing fish production, expensive, and inappropriate fisheries management. Further, stock enhancement efforts may undermine the genetic integrity of wild stocks at a time when limited resources might be better directed towards mitigating habitat degradation and/or reducing fishing effort. Although stock enhancement doesn't always make sense in ecological or economic terms, it remains popular because it represents a highly visible, job-creating, tangible effort to preserve employment in fishing communities. Using the terminology of the United Nations Food and Agriculture Organization (FAO), "culture-based fisheries" refers to stocking activities where aquaculture-reared animals are released into the wild and provide the basis for capture fisheries.

The opposite activity, where wild animals are captured and used as the basis for aquaculture, to be grown out (fattened) to marketable size in aquaculture facilities, is referred to as "capture-based aquaculture."[129] This type of aquaculture can be conducted in developing countries as well as industrialized countries, and represents an opportunity to farm species that fishery scientists cannot breed in captivity, on a commercial scale. The wild "seed" captured for grow out ranges from early juvenile stages to small adults. Not surprisingly, a significant portion (about a fifth) of aquaculture production depends on capture fisheries for their seed source. Capture-based aquaculture is practiced around the world, using many types of seafood, including, crabs, oysters, mussels, shrimp, and fish (e.g. eels, grouper, milkfish, tunas, and yellowtail). The largest species to be captured and fattened is also one of the most expensive species in the world: bluefin tuna, destined for the Japanese sushi market. Tuna farming started in Nova Scotia in 1975, when emaciated, tuna that had just arrived from their spawning grounds in the Gulf of Mexico, were fattened in holding pens prior to being shipped to Japan.[130] Within the Mediterranean, tuna farming started in Spain in 1995. By 2001, it had expanded to Croatia, Malta, and Italy, and by 2007, seven additional countries were farming or fattening Mediterranean tuna: Cyprus, Greece, Libya, Morocco, Portugal, Tunisia, and Turkey.[131] Farming tuna is also a big industry in Australia where it started in 1991, with Port Lincoln, in South Australia, the heart of the industry.[132] Beyond depriving wild populations of seed stock, this type of aquaculture frequently uses trash fish (or raw fish) as feed. This practice depletes local stocks and may result in fish-to-fish transmission of disease. Even when prepared feeds are used, they are derived from industrial fisheries. Thus capture-based aquacultural practices heavily impact wild fish populations both in terms of feed and seed. In the case of bluefin tuna, so many fish have been diverted into farming, that catches of wild fish have dropped substantially. With spawning stocks continuing to decline, and fishing mortality increasing rapidly, a collapse of some wild tuna stocks is quite possible.

Additional ecological costs of aquaculture include localized spread of diseases, the introduction of non-native species, the introduction of parasites, and the destruction of natural habitats and vegetation.[133] In tropical and sub-tropical regions, mangrove forests are very productive, ecologically rich, physically complex, stable, coastal habitats that are essential in the life history of many marine species. Despite the crucial role they play as prime nursery grounds, providing abundant food and shelter, for a variety of marine species, mangrove forests continue to be destroyed in order to accommodate the construction of ponds for shrimp farming. In addition to damaging coastal ecosystems, aquaculture often steals resources from wild populations.

Which is better, farm-raised or wild shrimp?

Considering whether to eat wild-caught or farmed seafood is not always an easy decision. As we stand in front of the freezer section of the super market, facing two packages of frozen shrimp, one labeled "wild," the other labeled "farm-raised," which should we choose? Should we support the production of farmed or wild seafood? Each has its own drawbacks. In a global context, most trawl fisheries for shrimp are extremely wasteful. The small mesh-size of their nets is responsible for excessive bycatch of non-target species. Frequently, as little as ten percent of their catch is shrimp. Although turtles may now be excluded from capture as the result of incorporating turtle excluding devices (TEDs) into shrimp trawls, the fish component of the bycatch, which is frequently discarded, often consists largely of juveniles of commercially important species. On the other hand, the global impact of shrimp aquaculture stems from the construction of shrimp ponds where productive, coastal mangrove forests have been razed, and the use of specially-formulated, high protein feeds that are produced by rendering distant, pelagic fish populations into fish meal and fish protein concentrates. The environmental impact of shrimp farms can extend well beyond the bulldozing phase. Once ponds or tanks are built, ground water and seawater are pumped in, then treated with chemicals and pesticides. As the polluted waste water is flushed out, it contaminates everything downstream to the sea. Salt water seeping out of shrimp ponds into groundwater ruins drinking water, and agricultural lands. Additionally, shrimp grown in high density aquaculture facilities are highly vulnerable to pathogen-borne diseases. The choice between wild-caught and farm-raised shrimp may seem like a conundrum. However, both American-caught and American-farmed shrimp are more regulated, in terms of environmental impacts, than imported shrimp, and as a result they may be a better ecological choice than imports. Furthermore, American shrimp are much less likely than imported shrimp to have been treated with drugs and chemicals that are not approved by the FDA. My preference is to consume American wild-caught shrimp, with the caveat: I eat shrimp infrequently.

Smaller species of cold-water shrimp such as pink (*Pandalus jordani*) and northern (*Pandalus borealis*) shrimp, that are typically sold as cocktail or salad shrimp, represent a good ecological alternative to large, warm-water, tropical shrimp such as tiger prawns (*Penaeus monodon*). Pink shrimp trawls utilize bycatch reduction devices that have reduced bycatch to acceptably low levels, of approximately 7.5 percent of the total catch.[134] Additionally, the height of the fishing line on the semi-pelagic, high-rise box trawls that are used for pink shrimp ranges from 12 to 30 inches (35 to 70 cm) above the bottom.[135] This type of gear produces less damage to the sea floor than bottom trawls. Spot prawns (*Pandalus platyceros*) caught in the north Pacific (particularly off British Columbia), are another ecologically sound alternative to warm-water shrimp. The fishery has very little bycatch and does little damage to the sea floor because spot prawns are caught with traps, rather than with bottom trawls.

Is farmed salmon as good as, or good for wild salmon?

In an effort to try to improve salmon catches, hatcheries have been aggressively used for decades (through governmental programs) to "supplement" wild salmon runs and mitigate the environmental consequences of constructing massive dams on salmon rivers such as the Columbia, in Oregon and Washington. During the golden age of dam building (1930s to 1950s), engineers viewed rivers principally as untapped sources of hydro-electric power, irrigation water, and routes for barge transportation of farm products and goods. Unfortunately, during the era of dam construction in the Pacific Northwest, salmon were largely an afterthought. The five North American Pacific salmon species, chum or keta salmon, pink or humpback salmon, sockeye or red salmon, Chinook or king salmon, and coho or silver salmon, are anadromous species that spend their adult life at sea, but return to their home rivers only once to spawn, then die. Many major dams in salmon watersheds were built without fish ladders, completely disregarding the needs of wild salmon, and cutting off their ability to return to their natal waters.[136] More recently, hatcheries have been used to produce young fish to grow out in net pens (through commercial enterprises). Stocks of wild salmon from different rivers are genetically quite distinct. Natural selection has fine-tuned the genetic make-up of wild fish to be an excellent match with the conditions in their native rivers. The genetic traits that make a good wild fish are not the traits that make a good farmed fish. In aquaculture, breeders aim for fast growing fish, not fish that can leap waterfalls and swim up fast-moving rivers.

Wild salmon populations face many problems from farmed fish. Escapees from aquaculture can interbreed with wild salmon of the same species, potentially contaminating the wild gene pool, and diluting the fitness of wild stocks. In traditional hatcheries, if wild eggs are collected, the collections are usually made at a limited number of sites, at times that are convenient for hatchery employees. This type of collection process limits the genetic variability of offspring, and may select for traits that are not important to the fitness of wild fish. Once hatchery brood stocks are established, multiple generations of hatchery fish yield inbred populations. The loss of genetic fitness that occurs in traditional hatcheries results in the loss of behaviors that are vital to wild fish, including courtship, and the ability to dig nests for their eggs. Farm-reared young salmon may outcompete wild young for food in rivers, but once in the sea, wild young survive better.

In another member of the salmon family, the steelhead (*Oncorhynchus mykiss*), a sea-run rainbow trout, new evidence has demonstrated that fish bred in captivity, in traditional hatcheries, for several generations have lost the ability to produce future generations of breeding adults.[137] This provides clear evidence that traditional hatchery-bred salmon of hatchery-stock origin are not biologically equivalent to wild fish. However, native populations appear to benefit when wild eggs and sperm are collected, eggs are fertilized, incubated, and hatched, then grown in a supplemental hatchery for a brief time period.

At the present time, it appears that net-pen-based farming does not relieve pressure on wild salmon stocks. In the Pacific Northwest, salmon farming has been banned in Alaska, hasn't grown beyond a small industry in Washington, but has become a large industry that exceeds the tonnage of the wild salmon capture fishery in British Columbia.[138] Wild salmon populations are stressed when the most frequently farmed species, a competing, non-native species, Atlantic salmon (*Salmo salar*), escapes from net pens. Though escapement may be downplayed, pen damage occurs due to storms, tides, accidents, and marine mammal predators, and can result in the release of thousands of farmed fish. Unfortunately, in the last decade, hundreds of thousands of Atlantic salmon have escaped net pens off Washington state and British Columbia, at a variety of life stages. Spawning of Atlantic salmon has been documented in several rivers in British Columbia.[139] Atlantic salmon appears to be an established, introduced species in the Pacific Northwest. Although it is extremely unlikely that Atlantic salmon will eventually replace native Pacific salmon species, by breeding in the wild they exert pressure on native fish, competing for resources, during every life stage.

Another problem for wild salmon is that net-pen facilities are frequently placed along migratory routes that young fish must travel en route to the sea. As a result, young fish sometimes have to swim past grow-out pens. In doing so, they often pick up heavy body burdens of the parasites and diseases that plague grow-out pens. Under normal circumstances young and adult salmon do not occur together in rivers or bays, so the young fish are not normally exposed to the parasites of adult salmon until they are bigger, and more capable of surviving an infestation. Parasitic copepods (sea lice) might not sound like a serious problem, but when juvenile salmon as small as 2 inches (50 mm) in length obtain a heavy body burden of sea lice after swimming past an aquaculture grow-out pen, their survival is greatly reduced.[140] Juveniles, lacking the fully developed complement of scales that protect the skin of adult salmon, are particularly vulnerable to parasitic sea lice and ensuing bacterial infections. One juvenile stage, the smolt stage, is a highly sensitive, transitional phase for salmon, the phase at which they leave fresh water and move into brackish estuaries on their way to the sea. Their osmotic balance changes completely, beginning with the challenge of keeping fresh water from diluting their saltier tissues and progressing to the challenge of stopping salt water from pulling fresher water out of their tissues. Facing this huge environmental challenge, smolts are not really prepared to cope with the additional physical challenge of parasitic sea lice.

The flesh of farmed salmon contains higher levels of contamination such as PCBs and organo-chlorine pesticides than wild fish. This is the result of the contaminants being more concentrated in industrial fish-meal feeds than in the prey of wild salmon. A comparison of the levels of persistent, bio-magnified contaminants in farm-raised Atlantic salmon has shown that European-reared salmon have significantly higher contaminant loads than salmon reared in North or South America.[141] The contaminant load in farmed salmon is most likely directly related to the contaminant concentrations in their feed, which have been shown to vary by twenty fold. Until the feeds used for farmed fish are decontaminated, ingesting farmed fish may pose health risks that undermine the beneficial effects of eating fish.

The flavor, texture, and color of wild salmon are generally considered to be superior to, and easily discernible from, farmed salmon. Chemical food coloring (synthetically produced carotenoid pigments) must be added to the feed of farmed salmon to turn their flesh from gray into the red hues of wild salmon flesh, which are obtained from crustaceans in their wild diet or food chain.

An alternative to net-pen salmon farming is being attempted in British Columbia, by the Coastal Alliance for Aquaculture Reform. Closed containment systems are land-based salmon farm systems that eliminate or greatly reduce 1) water pollution from excess feed, fish wastes, and chemical treatments, 2) fish escapes from the facility, 3) deaths of marine mammals and birds from shooting or drowning, and 4) the risk of disease transmission to wild fish.[142] They are also working to develop a more sustainable feed.

What about open-sea fish farming?

The future of salmon aquaculture may include large-scale movement of net pens into open seas or to partly protected channels with exceptional current flow. This type of location would likely minimize parasite problems, and the costs to control them, because sea lice do not adhere well to the skin of salmon in the strong currents of the open ocean, and the net pens would no longer be in the proximity of vulnerable wild smolts. Inside the pens, salmon would be fitter and firmer, because they would have to swim against the current, and this improved product quality could result in better prices. Organic wastes would be rapidly dispersed. However, moving net pens into more exposed conditions is expensive, and requires additional technology to deal with the problems of a remote site in terms of transportation, feeding, and maintenance, especially in stormy weather. As I write this, the Pacific Ocean off Oregon is calming down from storm-induced 40 foot seas, and nearly 100 mph winds. High winds and high seas may not happen often, but they certainly could wreak havoc with open-sea net pens.

Although salmon aquaculture is much larger and more widespread than other marine fish culture, many obstacles must be overcome before a wholesale switch into open-sea salmon farming would be possible. However, learning from the ecological impacts made by large-scale salmon aquaculture in shallow, protected coastal waters, significant progress has been made towards cleaner, greener aquaculture, by growing native species in submerged sea cages that are situated in deep ocean settings with strong current flushing. As of 2010, there had been only four examples of commercial open-ocean aquaculture in American waters.[143] Three offshore farms grew fish: one off Puerto Rico and two off Hawaii; and one grew shellfish off New Hampshire.

Cobia

Open-sea farming of cobia (*Rachycentron canadum*, the only member of the family Rachycentridae), was launched off Culebra, Puerto Rico, by Snapperfarm, Inc., in 2002.[144] Research in Puerto Rico was aided by expertise from the University of Miami and funding from NOAA grants. Snapperfarm's goal of was to start with a small-scale commercial enterprise, then expand operations to Panama. After six years in Culebra, they moved to Panama. As Open Blue, their first Panamanian cobia harvest occurred in 2010.[145] Cobia was chosen because it grows rapidly (ten times faster than some other species), has white flesh, a mild flavor, and appealing texture. Their product was marketed as Culebran Cobia (from Puerto Rico), and as Open Blue Cobia (from Panama). Hundreds of tons of cobia have being harvested from submerged cages at a depth of 220 feet (67 m), eight miles (13 km) off the Caribbean shore of Costa Arriba, Panama, over a flat, featureless bottom. The fish are raised in a clean, deep ocean environment with strong currents replacing all the water in the cages every few minutes. Utilizing the best of modern technology, the fish are fed a specially-formulated feed, from the surface, while they are being observed via a remote video camera to avoid overfeeding.

Moi

Hawaii has a long history (at least six hundred years) of aquaculture.[146] During the period when Hawaii was a kingdom, moi (*Polydactylus sexfilis*, a fish in the threadfin family, Polynemidae) also known as Pacific threadfin, was considered a royal food, and was raised in traditional Hawaiian coastal fish ponds, to feed the ali'i (the chiefs or highest class of Hawaiians). Moi is a white-fleshed fish, with a delicate taste and texture. Since 2001, moi have been grown in submerged, offshore cages, two miles (3.2 km) off Ewa Beach, Oahu, to nourish anyone who appreciates the traditional royal Hawaiian delicacy. Inside the cages, moi are fed prepared feeds that are 50 percent protein, half of which is derived from fish-based protein, and half of which is derived from plant-based protein sources. Cates International Inc., essentially took over a demonstration project (the Hawaii Open-Ocean Aquaculture Demonstration Program)[147] that was conducted by the Oceanic Institute (OI, an affiliate of Hawaii Pacific University) and the University of Hawaii, and in 2001 Cates received the first open-ocean lease for commercial aquaculture in the US. After permits were obtained, the project grew larger, although stock-enhancement hatchery facilities at OI continued to provide their fingerlings. In 2007, controlling interest in the company was sold, and the name was changed to Hukilau Foods.[148] After a few productive years, the long wait for permits to expand their facilities seemed to push Hukilau into bankruptcy, which they filed for in November 2010.[149] Despite bankruptcy reorganization, as of May 2011, Hukilau Foods was moving forward with plans to build a $1.5 million hatchery, and get back into moi production.[150]

Hawaiian yellowtail

Hawaiian yellowtail (*Seriola rivoliana*), also known as kahala, Almaco jack, Almaco amberjack, longfin amberjack, and Pacific amberjack (a large fish in the jack and pompano family, Carangidae), has been grown successfully in submerged, open-sea pens, above deep water, off the Kona coast, in Hawaii. In the wild, this species is prone to ciguatera fish poisoning. As a result, people are not interested in eating wild Hawaiian yellowtail. As previously discussed, ciguatera is a biotoxin that fish pick up through food chains that include herbivorous species that graze on coral reefs. By controlling what fish eat from hatchery through harvest, farmed yellowtail were raised ciguatera-free. After three years of research and development, in 2004, Kona Blue Water Farms, Inc. moved into commercial production. They utilized an integrated hatchery and offshore fish farm to produce sushi-grade fish, that was harvested on demand.[151] Trademarked as Kona Kampachi, this high-value seafood (a relative of the Japanese yellowtail or hamachi) was marketed as sustainably delicious, nutritious, and free from detectable levels of mercury and PCBs. The rich flavor, firm texture, and high fat content of farmed Hawaiian yellowtail made it suitable to serve either raw or cooked. The fish were sold in Hawaii, on the US mainland – from the west coast to the east coast, and in Japan.

Kona Blue Water Farms prided itself on being an example of environmentally sustainable aquaculture, nurturing their farmed Hawaiian yellowtail from the egg stage through harvest. Their hatchery did not deplete wild fish stocks via capturing wild fingerlings. Instead they spawned their own brood stock, and raised fingerlings in their hatchery, before transferring them to open-sea pens. During the grow-out phase, fish were fed high-protein feeds that utilized approximately equal portions of renewable plant-based proteins and fish meal/fish oil derived from Peruvian anchovies, which they considered to be a sustainable fishery. Thus, these feeds were more sustainable than traditional aquaculture feeds. To further minimize their impact on small, wild fish populations, in 2008, Kona Blue reduced the contribution of anchovy meal in their feed pellets to 30 percent, and made up for the lost protein by increasing soybean meal and adding chicken oil.[152] Environmental monitoring showed that the Kona Blue open-sea pens had negligible impacts on adjacent marine habitats. The submerged pens were located in a pristine, open-ocean site, offshore of Kona, Hawaii, over a sandy bottom, at depths up to 200 feet (61 m), where strong ocean currents flushed away excess feed and fecal matter. The geology and oceanography of Hawaii makes the siting of grow-out pens, a crucial factor in the success experienced by Kona Blue, difficult to duplicate outside of Hawaii. Three features make Hawaii ideal for off-shore aquaculture: deep water, strong currents, and warm water.

With grant support from the National Science Foundation and other groups, in 2011, Kona Blue Water Farms moved growout trials into a globe-shaped drifter net pen, named the "Velella" project, that floated freely in ocean current, 3 to 75 miles offshore, tethered to a manned sail boat. Hawaiian yellowtail grew to harvest size in six months, and were successfully harvested to some acclaim.[153] But in turbulent economic times, despite the success experienced earlier, Kona Blue Water Farms appears to have hit a rough patch. Kona Blue Water Farms was officially dissolved in November 2011. After a two-year transition, on June 4, 2012, Blue Ocean Mariculture, announced their acquisition of the former Kona Blue Water Farms, its hatchery, and offshore mariculture facilities.[154] The fish that they raise has been re-branded as Hawaiian Kampachi. As of 2012, much of their grow-out production had been moved to Baja California, Mexico, and they were installing new predator-resistant cages designed to reduce escapement, bio-fouling, and parasite burdens.[155]

Bureaucratic red tape and time consuming, expensive, legal entanglements in acquiring the permits necessary to increase production have hampered the expansion of offshore fish aquaculture startup companies in US waters.[156] As of 2007, there were 12 major federal statutes, administered by 7 federal agencies that affected marine aquaculture in US waters.[157]

Mussels

In 2007, America's first commercial offshore mussel farm was established off Hampton, New Hampshire by A. E. Lang Fisheries, in conjunction with the University of New Hampshire Atlantic Marine Aquaculture Center.[158] Their rope-cultured, blue mussels (*Mytilus edulis*, a bivalve mollusc), grown near the Isles of Shoals in the Gulf of Maine, have been marketed as "Isles of Shoals Supremes." The aquaculture begins when wild mussel larvae settle onto seed collectors. These seeds are later planted onto grow-out ropes which are encased in mesh socks. The seeded grow-out lines are suspended from submerged, anchored, backbone lines, with submersible floats used to adjust the buoyancy, as needed. After a growing season, the ropes are pulled out of the water to harvest the mussels.[159] Because the offshore environment is less turbulent than the near shore zone where other mussels have been farmed, offshore-raised mussels have thinner shells, and meatier bodies.

The concept of fish farming in the open sea is not without controversy. Advocates see it as a way to harness the blue expanses of the ocean, and have more control over the seafood we eat. Fishermen see it as a step towards privatizing ocean resources, to the detriment of wild fisheries. Areas of concern about open-ocean aquaculture range from species selection and technological issues, to social and economic impacts on commercial fisheries, and possible environmental impacts.[160]

Is any aquaculture sustainable?

Aquaculture definitely will continue to contribute significantly to the worldwide production of seafood in the future. However, the industry needs to shift towards more sustainable practices that minimize impacts on wild fish and shellfish populations. The rapid growth and development of aquaculture in the last few decades underscore the need for a cautious approach. At this point, the long-term consequences of aquaculture on wild populations, including disease transmission, genetic contamination, competition, and predation, are not clearly known, and are essentially inestimable. In order to address imminent and future global demands for seafood, the aquaculture industry needs to promote ecologically responsible practices that increase efficiency and environmental accountability, not just productivity, as well as reduce the negative impacts of aquaculture on wild fish and shellfish populations. Expanding seafood production through aquaculture may be inevitable. However, reducing the ecological footprint of aquacultural practices will facilitate an acceptance of farmed seafood by a public that wants to eat heart-healthy seafood but is concerned about the global consequences of doing so.

Traditional, extensive, aquaculture practices that require minimal supplementation may be the route towards achieving sustainable, or ecological, aquaculture. Inland aquaculture that is integrated with traditional farming methods can result in good yields of both fish and farm crops. Chinese have practiced multi-species integrated aquaculture in fish ponds in conjunction with rice farming for thousands of years. These practices are typically simple, inexpensive, low-maintenance, and rely on naturally available food resources, or farm by-products. Although the low densities of fish grown in extensive aquaculture result in low yields, these yields may be sustainable from one generation of farmers to the next. Classical Chinese carp pond culture has perfected the integration of aquaculture and agriculture. For example, five or more species of carp, that have slightly different feeding habits, can be raised together in ponds to efficiently utilize available food resources. Grass carp (*Ctenopharyngodon idella*) forage high in the water, on vegetation tops; bighead carp (*Aristichthys nobilis*) feed in midwater on zooplankton; also in midwater, silver carp (*Hypophthalmichtys molitrix*) feed on phytoplankton; mud carp (*Cirrhinus molitorella*) and common carp (*Cyprinus carpio*) both feed at the pond bottom, on benthic animals, and detritus (including carp feces); and black carp (*Mylopharyngodon piceus*) feed on molluscs.[161]

As marine fisheries decline, inland fisheries and aquaculture may become more important. This comes at a time when many inland waterways are facing environmental degradation from pollution, sediment loads, and the other physical and chemical costs of development and progress.

Seafood species that don't require high-protein, animal-based feed represent the greatest opportunity for sustainable aquaculture. In this regard, filter feeding shellfish, herbivorous (plant-eating) fish, or omnivorous species (all consuming, feeding on plants or animals) seem like the best candidates for sustainable aquaculture. Species that can be grown sustainably in integrated aquaculture systems include oysters, catfish, crawfish, and tilapia. Some trout species may also be grown in an ecologically sensitive manner.

Supporters of aquaculture point out that capture fisheries are not without collateral costs to other wild fish populations. They argue that many fishing techniques use bait in order to catch wild fish, including longline fishing, and using fish traps. In this regard, capture fisheries are frequently guilty of using wild-caught bait in order to attract larger target fish species. However, this only represents one meal in the life of the target species, as opposed to every meal of every day in the lives of farm-raised species. Capture fisheries are also criticized as being more damaging to ocean ecosystems than aquaculture, especially with regard to discarded bycatch of non-target species.

The one example of offshore aquaculture that has the best chance of becoming ecologically sustainable over time is offshore mussel culture. And the primary reason is because mussels feed themselves, by filtering the seawater that flows by. Small, wild fish are never ground up to feed them. Mussels clean the water, rather than pollute it. However, wild mussel larvae were collected. That represents a small negative effect on recruitment in wild populations.

Although aquaculture yields can provide a variety of benefits both to ecosystems and humans, all types of aquaculture negatively impact surrounding ecosystems to some extent. As a rule, truly environmentally responsible aquaculture remains elusive, and a work in progress.[162]

The Marine Aquaculture Task Force has made several recommendations to minimize environmental impacts, as follows. Marine aquaculture should be limited to native species. Care should be taken so that the genetics of wild populations are not diluted by escapees from aquaculture. Aquacultural practices should minimize the likelihood and impact of diseases and parasites, and they should not degrade water quality or the health of marine ecosystems. And only sustainable feed ingredients should be used.[163]

Can we trust imported seafood?

About half of the seafood imported into the US is produced by aquaculture. Farmed fish and shellfish that are raised in high-density culture facilities are extremely vulnerable to diseases and bacterial infections. Although the US has stringent regulations that specify which drugs and chemical treatments are approved for use in aquaculture, other countries have different standards, and may lack enforcement. In the US, the FDA is in charge of ensuring the safety of imported seafood, with regard to residues of unapproved drugs or treatments. However, its oversight is limited in scope, especially in comparison to the comprehensive reviews the European Union (EU) conducts on the food safety systems of countries that export seafood to Europe. FDA sampling has been both inadequate, with regard to the number of drug residues that it tests for (less than half the number tested for by the EU), and sparse. For example, in 2009, FDA tested only about 0.1 percent of imported seafood for drug residues.[164]

Meager sampling of imported seafood and the fact that a number of countries that export seafood, including Bangladesh, Chile, China, Indonesia, and Vietnam, permit the use of drugs that are unapproved for use in aquaculture in the US, combine to make eating imported, farmed seafood a disconcerting, disquieting experience. Many countries have exported seafood to the US that contained residues of drugs that are not FDA-approved for use on farmed seafood. Producers that are found to be out of compliance with seafood regulations are placed on the FDA's import alert list, and their seafood products are detained without inspection. As of September 25, 2012, fifty eight seafood producers in China, Malaysia, Mexico, Taiwan, Thailand, and Vietnam remained on the import alert list.[165] Over forty percent of the Chinese producers on the list were cited for adulterated tilapia. Government inspections of drug residues in imported seafood that were conducted by the EU, US, Canada, and Japan, between 2000 and 2009, found that Vietnam had the most violations, and China was second.[166] Shrimp and prawns most frequently exceeded drug residue limits. China is the leading supplier of seafood to the US, especially farmed-raised shrimp and tilapia. In June 2007, after eight months of increased monitoring revealed that 25 percent of the Chinese farmed seafood samples contained disallowed chemicals, the FDA placed import restrictions on shipments of five types of farm-raised seafood from China: shrimp, catfish, basa (in the catfish family), dace (in the carp family), and eel.[167] In some cases the contaminants were anti-fungal treatments, such as gentian violet and malachite green, and in other instances, the contaminants were antibiotic agents such as nitrofurans and chloramphenicol.[168] Exposure to some of these treatments, such as malachite green, a known carcinogen, can result in an increased risk of cancer. But in the case of antibiotics, a far greater concern is that the broad, indiscriminate, overuse of some of our most powerful antibiotics, such as ciprofloxacin, could hasten the development of drug resistant microbes. This would weaken the ability of these "super" drugs to fight human infections. Firms that have been caught shipping adulterated seafood products to the US are placed on an import alert, which results in detention of all subsequent shipments, until the firms prove their compliance with US regulations.

In our global village, one nation has recently been receiving a lot of bad press regarding product recalls due to contamination. That nation is China. The contamination has not been limited to the lead paint used to give millions of children's toys a shiny appearance. Rather, contamination has also been detected in food products. In 2007, melamine was added to pet food products, in order to boost their apparent protein content. This resulted in the deaths of scores of beloved pets in the US that ingested pet food made with melamine-laced, Chinese rice protein concentrate or wheat gluten. Chemically, protein and melamine are both nitrogen-rich compounds. Unfortunately, the protein content of food products is often tested by simply measuring nitrogen content, so melamine was used as a cheap counterfeit protein. In September 2008, the problem of melamine contamination moved into human food products, when several Chinese dairy products, especially baby formula, were discovered to be contaminated with melamine, after thousands of Chinese children were sickened with kidney stones, and several died from kidney failure. Unmitigated, capitalistic greed was the likely reason that dairy products intended for consumption by children were adulterated, as the addition of melamine could either boost the apparent protein content of poor-quality milk or mask the fact that manufacturers were fraudulently diluting milk. Although the problem was known earlier, the Chinese delayed the release of news of the melamine-contaminated milk until just after the Beijing Olympics.[169] Hot on the heels of that news story, in October 2008, food-safety inspectors in Hong Kong discovered that Chinese-grown eggs were contaminated with melamine. The source of the melamine contamination was determined to be chicken feed that was adulterated through the addition of counterfeit protein, i.e., melamine in powder form. Chickens that ingested "high-protein," melamine-contaminated feed laid the contaminated eggs. The appearance of melamine in animal feeds raises new concern that other farmed animals that are fed Chinese-made processed feeds (or feeds with Chinese-made wheat gluten or rice protein components), such as fish, shrimp, and hogs, may also be vulnerable to melamine contamination. Each new report of melamine contamination further taints the image of food bearing the "made-in-China" label.

China produces and exports more seafood than any other country in the world. Chinese dominance in fishery production, and doubts about the accuracy of data prior to China revising their fishery statistics in 2006, have led the UN Food and Agriculture Organization (FAO) to present some of their global fishery statistics simply as "China" and "world excluding China." Seafood production from Chinese aquaculture exceeds their capture fisheries. More than half of the world's aquaculture production comes from China. The bulk of this production comes from China's immense inland fishery. Unfortunately, much of their inland waters, where fish are cultured, are heavily contaminated with municipal and industrial effluents (including untreated sewage), as well as farm runoff.[170] Chinese coal-fired power plants that spew mercury emissions also contribute to contamination of inland waters. In China, industrial development has recently charged ahead of environmental quality concerns. In order to maximize production, intensive aquaculture practices are used. High density stocking results in over-crowded ponds, and stresses the enclosed fish populations. Poor water quality produces an additional stress on the fish. Farmers treat their ponds with drugs and/or antibiotics to keep the fish alive, and to keep disease from spreading in their ponds. Many coastal waters in China, and the fish grown in them, are contaminated with pesticides and other toxic wastes. A recent analysis of seafood obtained at fish markets in Guangdong Province discovered alarmingly high levels of organochlorine pesticides in shellfish.[171] Despite Chinese bans on the use and production of some of the pesticides DDT and HCH since 1983, they persist in the environment. High level of DDT and HCH were detected in water, sediments and seafood products in coastal southern China, and some of this may be from recent usage of DDT. This region may be one of the most DDT-polluted areas in the world.

Regardless of how hygienic fish processing facilities and procedures may be, fish and shellfish that are contaminated with drug residues, antibiotics, heavy metals, or pesticides, cannot be cleaned up. The safety of seafood reared in Chinese aquaculture raises serious concern, internationally and domestically. Some Chinese consumers prefer to eat marine fish species that were collected using destructive blast-fishing techniques rather than eat anything caught or raised in their polluted inland waters.

Consumers who want to choose seafood wisely and identify its source can benefit from reading country-of-origin labeling on fish and shellfish, prior to purchase. Since 2005, the US Department of Agriculture has required supermarkets to label where seafood comes from and whether it was wild-caught or farm-raised.

Will world demand for seafood exceed world seafood production?

Over seven billion people now inhabit planet Earth, and our population is growing at rate of nearly 100 million people a year. As more people embrace the remarkable health benefits afforded them by including seafood as a major component of their diet, the demand for seafood could quite possibly exceed the supply, the maximum sustainable worldwide production of seafood, in the not too distant future. In the short run, in developed countries this would raise the price and reduce the availability of seafood, and in developing countries this would reduce the food security of millions of poor people who depend on cheap fish as their primary source of protein. It would also put additional pressure on the remaining fishery stocks.

One of the best sources of information on fishery stocks is *The State of the World's Fisheries and Aquaculture*, also known as SOFIA, published biennially by the FAO Fisheries and Aquaculture Department. The following data were obtained from the most recent SOFIA publication.[172] Global marine capture-fishery production has been nearly level since 1985, at around 80 million tonnes (i.e., metric tons). Although tonnage has remained about the same, the composition of the catch has varied substantially from year to year, especially for small pelagic species, such as populations of Peruvian anchoveta, in response to El Niño conditions. Between 1985 and 2010, inland capture-fishery production increased from about 6 to 11 million tonnes. At the same time, fishery production from aquaculture has grown significantly. Global food fish production in aquaculture in 2010 (at 59.9 million tonnes) was twelve times greater than in 1980. All together, marine and inland capture fisheries and aquacultural production, produced 148.5 million tonnes of fish in 2010, of which 128.3 million tonnes were used for human food. During 2010, 54.8 million people were engaged in fish production as fishers and fish farmers, utilizing an estimated 4.36 million fishing vessels. Globally, seafood production is an important industry. The world's top six nations in aquaculture production in 2010 were all Asian, led by China, which accounted for 61.4 percent. The next five nations, India, Vietnam, Indonesia, Bangladesh, and Thailand, together contributed 20.4 percent of global aquaculture production. Since 1980, aquaculture has grown rapidly, and has driven the growth in world food fish production. Although feed is generally considered to be a major ecological and economic constraint to aquaculture, in 2010, species that were grown without added feed (i.e., filter-feeding bivalves, such as clams, oysters, and mussels, and plankton-feeding carp grown in polyculture systems) comprised one third of farmed seafood production. Unfortunately, the biggest growth in aquaculture production has been in fed species, and global non-fed production has gradually declined from 50 percent in 1980 to 33.3 percent in 2010.

SOFIA 2012 also presented a comparison of the status of world fishery stocks between 1974 and 2009. In 1974, 40 percent of fish stocks were classified as not fully exploited. By 2009, not-fully-exploited stocks had dropped to 13 percent. In 1974, 50 percent of fish stocks were considered fully exploited (or fully fished), while in 2009, 57 percent of fish stocks were fully fished. The remaining 10 percent of fish stocks in 1974, and 30 percent in 2009, were classified as overexploited or (overfished) fish stocks. Based on the latest numbers, no more than 13 percent of the world's fishery stocks are thought to be capable of producing more, in response to an increase in fishing pressure. However, fishery scientists are certainly not recommending increasing the fishing pressure on each and every not-fully-exploited stock.

The latest official fishery statistics (SOFIA 2012) provide limited data on the improvement (or recovery) of a few marine fishery stocks, however, most marine stocks are heavily exploited and are unlikely to significantly increase in the future. Additionally, FAO data show that marine production reached its peak in 1996, at 86.4 million tonnes, and gradually declined to 77.4 million tonnes in 2010. Doubting that global fish catches could continue to increase while the vast majority of fish stocks were being full fished or overfished, a pair of sleuthing fisheries ecologists, Reg

Watson and Daniel Pauly from the University of British Columbia, determined that China had over-reported fishery catches in the mid-1980s at least through 1998.[173] After correcting for this distortion and removing the chaos in production statistics caused by wildly fluctuating anchoveta stocks, they found that world fishery landings appear to have peaked in 1988. Regardless of when landings peaked, catches are declining, the percentage of overexploited fish stocks has increased, the percentage of not-fully-exploited fish species has decreased, and a global imbalance exists between fish stock levels and fishing capacity. Indubitably, the state of the world's marine fisheries appears to be worsening. The vast majority of fishery stocks are currently stressed by heavy fishing. Stressed populations are increasingly vulnerable to human-induced environmental perturbations such as global climate change and pollution, as well as naturally occurring oceanographic changes such as El Niño events, and decadal-scale oscillations in currents and climate. At this point, without improved fishery management and attention to overcapacity of the fishing fleet, we could easily be facing a serious decline in marine production.

In a somewhat controversial move, a group of international scientists have recently predicted that the world's seafood supply could crash and fail utterly by 2048, if species continue to decline at the current rate.[174] They defined the collapse of fisheries as the point when fish catches drop below 10 percent of their historic levels. Based on this definition, they reported that stocks of 29 percent of all fished species had collapsed by 2003. This has caused a lot of defensive denials by the seafood industry, and rebuttal of their modeling techniques by other fisheries scientists. Regardless of whether their prediction is true or not, it has sparked a necessary conversation about the potential collapse of seafood populations, worldwide, and created an appreciation of the fact that we need to make informed choices when we eat seafood.

How can we make better use of our seafood resources?

Conservation is an old concept, and yet it has recently been embraced as a very contemporary idea as well. Natural resources need to be utilized and managed in ways that protect them against excessive depletion. Globally, marine resources need to be conserved now to ensure that they will both exist and be productive in the future. Marine resource managers and environmentalists both work towards this goal, but in different ways. All levels of government as well as non-governmental organizations and international agencies are working towards the conservation of marine resources. However, governments frequently find themselves caught between conflicting goals of fishery economics and fishery science, and the right decisions are not always made. Decisions often favor jobs over resources and, indirectly, the future of both.

Despite years of warnings from fishermen that their catches showed all the classic signs of an overfished fishery, conflicting motivations of fishery scientists and politicians may have resulted in the permanent ruination of one of the most storied fisheries in the world, the northwest Atlantic cod fishery. The fabled history of this fishery has been richly documented by Mark Kurlansky in his book *Cod: A biography of the fish that changed the world*.[175] And its downward spiral, following 400 productive years, is compellingly told by Michael Harris in his book *Lament for an ocean: The collapse of the Atlantic cod fishery, a true crime story*.[176] Before adequate conservation measures were enacted, the breeding stocks of Atlantic cod were essentially fished to commercial extinction. And following more than 10 years of a moratorium on cod fishing, an ecological regime shift may have occurred. Off Newfoundland, the ecosystem that was once dominated by large fish, is now dominated by invertebrates: snow crabs and cold-water shrimp.[177] The cod may never return in sizable numbers.

Overfishing and mismanagement of fisheries have pushed fish stocks to the breaking point, to the end of the line. Our ability to exploit fishery resources has outpaced our ability to manage them. Modern technologies have given fishers an unfair advantage and have left fish stocks with virtually no place to hide. The world needs to fish smarter not bigger, deeper, longer, or harder. More intelligent fishery management is necessary. Only sustainable marine fisheries can endure.

Responsible, sustainable marine fisheries and healthy, dynamic ecosystems go hand in hand. The most successful form of sustainable fishery management will likely involve an ecosystem approach. Marine ecosystems provide a number of "services," products and processes that are beneficial to us. These services range from the biodegradation of waste water, moderation of climate, production of food and fresh water, generation of oxygen and absorption of carbon dioxide, to providing tourism and recreation opportunities.[178] Ecosystems that function well will continue to provide all the services that we expect.

Ecosystem-based fisheries management necessitates a precautionary approach, to ensure the integrity of the ecosystem, and take into account the uncertainties inherent in complex marine ecosystems. Whenever possible, management should err in favor of the ecosystem rather than the industry. Fishing efforts should be highly selective, with regulations focused on what may be caught, rather than what was caught. Discards (dumping unwanted catch components at sea, dead or alive), and bycatch (catching non-target species), should both be minimized, or better yet, eliminated. They don't just represent collateral damage, they represent the loss of balance, diversity, and functional biomass from the ecosystem.[179] Destructive fishing gears and methods need to be avoided in order to conserve structurally and biologically complex habitats. Reductions are necessary in both fishing capacity and fishing mortality. And as a result, fishing effort, the number of boats, and the number of fishermen should all be reduced.

One of the best ways to perpetuate diverse marine ecosystems is through setting aside intact habitats as no-fishing reserves. Marine protected areas (MPAs), large marine areas that are managed with the long-term goal of conserving marine natural resources, where human activities are regulated, need to be part of any solution to the current fisheries crisis. After MPAs have been established, fish production on adjacent fishing grounds has frequently increased dramatically, both in terms of fish size and abundance. Although many fishermen like the concept of MPAs, some fight to keep their traditional fishing grounds from being designated as a MPA. This behavior reflects a place attachment as well as a not-in-my-backyard attitude of those who actively support a concept, but do not want to be personally inconvenienced by its implementation.

Managing fisheries sustainably and establishing MPAs require much more than simply an ecological or oceanographic perspective. Human dimensions (social, economic, cultural, and institutional) also come into play.[180] Any transition towards sustainable fisheries will impact fishermen as well as all other local stake holders (users of the marine environment). Before MPAs are created, competing uses have to be addressed. For example, wave and wind energy and resource extraction facilities are now vying with fishers for access to inshore marine sites. Tourism, recreation, and sport fishing may also have conflicting interests.

The economics of fishing are complicated by the fact that fishing is a heavily subsidized industry. Fishery subsidies can be loosely defined as government policies that aid the industry, usually with financial benefits. They enable fishing to continue through economic conditions that would otherwise be unprofitable. A careful analysis of fishing subsidies has characterized them as the good, the bad, and the ugly.[181] Good (beneficial) subsidies enhance the growth of fishery stocks, and represent an investment in fishery resources. They include money spent on fishery management programs, research and development, and MPAs. Bad (capacity-enhancing) subsidies represent disinvestment in fishery resources. These include money spent to subsidize fuel, boat construction and modernization, fishing port development, price support, processing and storage infrastructure, fishery development projects, and foreign access agreements. Ugly (ambiguous) subsidies may lead to either investment or disinvestment in fishery resources. They include fisher assistance programs and vessel buyback programs. For a single year, the total of global subsidies for marine capture fisheries was estimated to be $27.2 billion. Due to heavy governmental subsidies, the true costs of fish are not reflected in the market prices that consumers pay. In order to move towards sustainability in world fisheries, capacity-enhancing subsidies need to be drastically reduced, or eliminated.[182] In order to transition from business-as-usual fishing to sustainable fishing, fishing capacity must be reduced. Many boats will need to be removed from the fleet, many fishermen will need to stop fishing. However, to achieve this, in the short term, financial support will be necessary to facilitate this transition, to redirect the employment of displaced fishers. Perhaps a portion of discontinued subsidies could be used to train fishers for other work.

There is abundant evidence to suggest that fish stocks will continue to decline unless we move towards sustainable marine fisheries. The journey could be slow, but it is the right path to take. Consumer demand for sustainably-caught seafood will bolster the movement to sustainable fisheries. Now, if we could only get adequate governance and enforcement to bring IUU (pirate) fishing under control, we might feel even more optimistic.

Can we switch to underutilized species?

While it is tempting to think that our oceans contain such bountiful food resources that fishermen can easily switch their efforts to a new species whenever they deplete the stocks of preferred species, as they have done for centuries, this is extremely unlikely to happen very often, at any time in the future. As more and more capital has been invested in fishing, boats have been equipped with very sophisticated electronic technologies that effectively annul spatial refugia for fish stocks and species. Captains can view the sea floor topography from the comfort of the wheelhouse. They can examine water temperatures, which often correlate with fish distribution. Through global positioning systems (GPS), they can utilize satellite triangulation to pinpoint exact geographical locations where fishing efforts have been productive, and return to the same fishing spot, whenever they like. Electronic fish-finders use active sonar to echo-locate targets, notably schools of fish and the bottom. Although not that long ago fish stocks may have had places to hide in the sea, they are now much less likely to elude their sophisticated electronic seekers. Some of the most recently developed fisheries have exploited long-lived, deep-sea species that inhabit seamounts in southern oceans. So, in the hide-and-seek game of fishing, the seekers have already searched whatever sea floor topography mirrors productive areas that they have already fished out.

In simpler times, fishing effort was measured in catch per unit effort, for example in pounds/kilograms of fish caught per man hour of fishing. At a time when fishing techniques were less sophisticated and fish stocks were larger, catches were directly related to the fishing effort. However, today's fishing effort is hardly comparable to the "man hours" of old, as electronic technologies now focus fishing effort. Prior to the development of fish-finder and GPS technologies, commercial fishermen would need to rely on their knowledge of the seasonally-changing biology and behavior of fish species, to predict when and where to fish. While that knowledge is still paramount to successful fishing, new technologies give modern fishermen an advantage in terms of where to fish.

The concept of switching to "underutilized" species reflects the fallacy of underutilization. Just because a fish is locally abundant, doesn't mean that it should be harvested for human consumption. Within the oceans, many small schooling fish species sustain populations of large predatory species. Before we catch all the bait fish, we need to consider the trophic significance of prey species in marine food webs. Although the term "underutilized" implies that greater exploitation of a species is desirable (as fisheries economists may suggest), many ecologists feel more comfortable replacing this term with "lightly fished" or describing the species as "subject to low fishing mortality." The reasons that species are lightly fished can vary widely. For example, an examination of British fish sales revealed a mass market preference for large fish, with firm, white flesh, that is mild flavored, something like the cod fish that provisioned them nobly for years. And British consumers generally want the fish that they purchase to be easy to prepare, without skin and bones. Thus, if sizable stocks of small fish are close to shore, British fishermen would not necessarily obtain a good price for a haul of them, because small fish don't have great fresh market appeal to their local populace. Small, bony fish species, such as sardines, that are destined to become human food frequently end up as a canned product. Their many bones are not an issue, because the heat processing of canning renders the bones edible, adding calcium to the fish's nutritive value. However, not every fishing port has the canneries necessary to process small fish. Building new canneries is capital intensive, and unquestionably a large investment in processing facilities does not ensure that sizable fish stocks will remain in the area. Abandoned fish canneries dot the west coast of the US. One of the most famous is Cannery Row in Monterey, California, immortalized by John Steinbeck in his book of the same name. In less than 40 years, Monterey went from being the "Sardine capital of the world" to a collapsed fishery.

Very small fish are not appealing to every market, nor are fish that are slimy, ugly, bizarrely-shaped, dark fleshed, or oddly named. However, if fish flesh is suitable in appearance, then ugly, slimy, and bizarrely-shaped fish can be skinned and butchered, and marketed as fillets, without the public ever knowing what the whole fish looked like. And oddly-named fish can be re-branded as something that sounds more appealing. This was done with the Patagonian toothfish which was successfully marketed as Chilean sea bass. Unfortunately, dark-fleshed fish encounter more resistance in a marketplace where customers have traditionally conservative tastes in fish. Some cultures are decidedly more adventurous in the seafood they consume. For example, within Europe, Italian, Portuguese, and Spanish cuisines are notable for their great appreciation of whole, dark-fleshed, strong-flavored fish in the round, as well as squid and other shellfish. Asian markets probably embrace the greatest diversity of seafood. Although Thai and Chinese cuisines feature a diverse array of foods from the sea, Japanese tastes and markets represent the pinnacle of appreciation of virtually everything edible from the sea. As an island nation, they have always looked to the sea for their food.

Historically, many stocks of soft-bodied fish were lightly fished partly because they required special handling. Although they represent a potentially rich protein source, the larger the catch, the bigger the problem. To some extent, fishermen can improve the quality of their catch of soft-bodied fish by decreasing the total volume of their catch. However, this concept is counter-intuitive to the maximize-the-catch instincts of most fishermen.

Can we utilize fish more efficiently?

Perhaps a better interpretation of "underutilization" should switch from searching for new target species to exploit, to thinking about how to most efficiently utilize the fish that are already being caught, both in terms of maximizing protein available for human consumption, and beyond. One of the factors that brought about the downfall of the Monterey sardine fishery was the fact that the abundant fish were rendered into oil and fishmeal as well as canned for human consumption.[183] Although canning required specific sizes of sardines, which represented mature fish, the rendering plants took all sizes, which quite likely included juvenile fish. Aggressively fishing juvenile stocks can rapidly diminish the size of any future catches. At a time when currents and water temperatures were changing, poor regulation of this super-abundant, sardine fishery ultimately destroyed it.

The size and shape of many fish species pose difficulties in processing the catch. Long ago, fish were all processed by hand, but now machines do most of the work. At present, many fish species can be mechanically gutted, skinned, and filleted, but not all fish species conform to the shape the machines were designed to efficiently process. Some fish require little more than beheading and gutting to be marketable. However, as fish are getting smaller, one trend in seafood technology is to form laminated blocks of fish fillets. These blocks are then frozen, and cut into uniform pieces suitable for marketing as fish sticks, fish fingers, and fish fillets. In most cases, fish products that have been cut from laminated blocks are battered and breaded in preparation for frying, at home or at any number of fast food restaurants.

Once fillets have been removed, some flesh remains on the carcass. Seafood technology has also come up with a way to harvest the odd bits of potentially-underutilized protein that remain. Carcasses can be passed through deboning machines that remove all soft tissue from the skin and bones, producing minced fish. Soft-bodied fish are well suited to pass through deboners and end up minced. To some extent, a small amount of minced fish can be added to laminated fish blocks without degrading the quality of the product. However this is not the primary use for minced fish.

Surimi

Following in the Japanese culinary tradition, minced fish is most often made into surimi, a Japanese term meaning ground fish or meat. All fish used for surimi are beheaded and gutted. The cleanest mince (of highest quality) is obtained when boneless, skinless fillets are run through the mincing machinery, as blood, connective tissue, and other contaminants were removed prior to mincing. The greatest yield of fish mince results when whole beheaded, gutted carcasses are run through the mincing, deboner machines, although the resulting mince is of lower quality. If the gut cavity is thoroughly washed, then fish mince yield can be maximized without sacrificing much quality when butterfly-shaped, skin-on, fish fillets are run through the mincing machinery. Regardless of which form of fish goes in to the mincing, deboner machines, the ensuing mince is washed several times (to remove fat, blood, etc.). Protein fibers from striated muscles (myofibrillar proteins) that form the main component of the surimi protein gel, constitute approximately two thirds of the minced fish meat.[184] The remaining third consists of undesirable components that are removed via washing to yield a colorless, odorless surimi. After washing, refining, and pressing, a small amount of sugar or another cryoprotectant (a chemical substance that protects the proteins from freezer damage) is added, to stabilize the myofibrillar proteins and to increase the shelf life of the frozen product. At this point, the uncooked surimi fish paste (a protein gel) becomes essentially a blank canvas for artificially-flavored seafood products, notably fake crab and lobster. Until recently, in the US, surimi-based crab, lobster, and scallop products had to be labeled "imitation." However, that term generated a negative impression with consumers and confusion as to whether the product was actually seafood or "imitation" seafood. As a result, FDA regulations were changed in 2006 to allow the less off-putting label description of "flavored seafood, made with surimi, a fully cooked fish protein." To some extent, the terms "surimi" and "kamaboko" are used interchangeably. However, surimi also describes the unfinished fish paste, while kamaboko refers to the finished product, which is fully cooked, usually steamed.

Surimi is processed at sea in factory ships or onshore, using abundant, mild-flavored, low-fat, white-fleshed fish, such as walleye pollock or Pacific hake (*Merluccius productus,* a member of the Merlucciidae family of cod-like fishes). Frozen surimi is then delivered to factories where it is processed into various forms of kamaboko, cooked seafood analogues that are ready-to-eat products, which are marketed to restaurants and the general public. The transition from surimi to kamaboko involves creating more complex texture through the addition of starches or proteins such as potatoes or egg whites.[185] The desired seafood flavors are created through the addition of flavorings, either natural or artificial, and sometimes both. In order to create the desired product appearance, colors are adjusted, sometimes through early-process bleaching and/or the late-process addition of artificial colorings. Once the product has the desired taste, texture, and color, it is then either extruded or molded into one of many shapes, and the product is cooked. While all kamaboko products can legitimately be labeled as seafood, they are definitely processed foods. And some of the nutritive value of the fish could be lost through various processing steps. Those of us who view fish as a healthful, tasty, wild source of protein might be have some misgivings about the almost Spam-like processing involved in making kamaboko products. However, although the sodium content of kamaboko is higher than wild fish, and the nutrient content of surimi-based seafood differs significantly from that of the corresponding real shellfish that it imitates, even considering all the additives that have gone into it, kamaboko is still a much healthier food choice than Spam or other processed meat products. In North America, one of the most widely available kamaboko products is simulated crab, also known as kanikama, short for kanikamaboko. California rolls often feature kanikama, and slices of it are frequently added to bowls of miso soup. In deciding to eat or not eat kamaboko, think about this: in North America kamaboko products are most often made from Alaska pollock, which is considered a sustainable fishery. So, in the ecological scheme of things, when we include surimi-based, kamaboko seafood products in our diet, we are making a fairly smart choice.

The same may be said for substituting one abundant fish for a more prized catch, which might be inaccessible or too expensive for a local market. For example, years ago seafood processing scientists realized that although canned chub mackerel (*Scomber japonicus*) is pinker and tastes more like salmon than Atlantic mackerel (*Scomber scombrus*) does, they both have a texture similar to salmon. This led to the suggestion that both chub and Atlantic mackerel might be transformed, via the skillful addition of artificial flavor and red pigment, in the canning process, to a suitable substitute for canned salmon. There is nothing nutritionally wrong with eating pigment-enhanced, salmon-flavored, canned mackerel, as long as the product is very clearly labeled as to what the can contains, and the consumer is not misled to believe that it is canned salmon.

One crucial and sobering loss in fishery production is the vast amount of fish that are wasted between the time that they are caught and the time that they are landed for processing, due to spoilage and infestation with insects, particularly in developing countries. In African artisanal fisheries, post-harvest losses have been estimated at 20 to 25 percent of the landed catch, with losses sometimes as high as 50 percent.[186] About 30 years ago, after being shown where Egyptian fishermen landed their catches on Lake Nasser, behind the Aswan High Dam on the Nile River, I was startled to see the poor condition of the landed fish. Unfortunately, none of the artisanal fishermen had any kind of refrigeration or ice to chill their catch. As a result, at that time, the loss due to spoilage was estimated at 50 percent. Many factors contribute to the spoilage of fish, including temperature, humidity, oxygen level, salinity, condition of fish before catch, a fish's body burden of microbes, the size, shape, and fat content of a fish, and the amount of time elapsed since capture.[187] Vanquishing spoilage losses in tropical, artisanal fisheries could simultaneously reduce overfishing on some fish stocks and reduce food insecurity in some developing countries.

At times, it seems that fishery problems are so vast that there isn't much we can do to change things. After all, how much can the responsible, sustainable behavior of one person impact world fisheries? Well, certainly when nothing is ventured, nothing is accomplished. We can think big, but we should start by taking small steps. For example, in the course of studying various aspects of the recruitment and biology of Nassau grouper, in the Bahamas, I came to realize how vulnerable all species of grouper are to overfishing. As a result, I knew that I couldn't eat grouper. And as a scuba diver, I have spent many hours enjoying the proximity of various grouper species. They are often curious, and allow divers to closely approach them. In Bonaire (Netherlands Antilles), an island in the Caribbean Sea, just north of Venezuela, defined as the "Diver's Paradise," by their license plates, several restaurants feature grouper on their menus. Divers are led to believe that the grouper is farm raised, and is not caught on the local reefs. However local fishermen are always interested to learn exactly where divers have seen large grouper, and then the grouper soon mysteriously disappear from the reefs. While I can't single-handedly stop people from eating grouper, I started a grass roots movement in Bonaire. Armed with a few tee shirts that featured an enhanced photo of a tiger grouper, in January and February 2010, I enlisted a group of friends to wear the messages "Enjoy Grouper on the Reef, Not on your Plate" and "Take Grouper off the Menu." Our band of grouper groupies wore the shirts around the island, including out to dinner, and received a lot of support and interest from fellow divers and restaurant staff. I shared my message with local, knowledgeable, advocates that I respected. One of these, Bruce Bowker, owner of Carib Inn, liked the message so much, that I gave him the design, so he could spread the word, which he did on a not-for-profit basis. As of May 2010, divers could buy the shirts in Bonaire, and then take the message with them when they return home to the US, Canada, Holland, or wherever they live. And as tee shirts often exemplify the preferred garb of tropical divers, they represent a great place to wear important messages supporting wild, healthy oceans.

We certainly need to have more respect for the marine resources that we rely upon to feed us. These finite resources require thoughtful management that results in conserving life on earth, not destroying it. Wise use of marine resources necessitates thinking outside of our normal perspective. This kind of ecological thinking harkens back to times when people were more connected to the world they lived in. Historically, coastal Atlantic populations that relied on cod fisheries, from the Basques to Nova Scotians, utilized the odd bits of the codfish: including heads, tripe, cheeks, tongue, and roe.[188] In a restaurant in Newfoundland, about 35 years ago, I ordered, and enjoyed dining on cod cheeks and tongue, eating like a local.

The indigenous approach

Around the Pacific rim, indigenous people have depended upon returning salmon populations to sustain them for hundreds (or maybe even thousands) of years. They have developed a reverence for the spirit of salmon. These deep connections have been captured in *First fish first people: salmon tales of the North Pacific rim*.[189] Although the tales have been told by people from many different cultures, together they reflect the importance of salmon in their lives and the complete use of salmon as a resource. Shiro Kayano, an Ainu from Hokkaido, Japan, described his father making traditional shoes from the strong, thick skins of salmon that had already spawned. Historically, to sustain themselves through the winter, the Ainu harvested salmon after it had spawned. Rather than jeopardize salmon runs by catching large numbers of fish before spawning, they chose to live off the interest of their natural world. Nadyezhda Duvan, an Ulchi from Siberia, wrote that Ulchis utilized every part of the salmon: the bladder was a source of glue; vertebrae and cheekbones were used to make children's rattles to ward off unwanted spirits; fish skins made rain-shedding boots and garments. Elisabeth Woody (Navajo/Warm Springs/Wasco/Yakama), from the Pacific Northwest, recalled eating all of the salmon, except its guts; drying heads and gills; with dried spines and tails eventually ending up in soup. Nora Marks Dauenhauer, a Tlingit from

Alaska, recounted dryfish camp from her childhood when her extended family gathered in the fall, to dry salmon in smokehouses. Her father's favorite treat was raw coho heads and tails. But the big salmon heads were used to make fermented fish heads. For this, a large pit was dug in the beach, on a low tide. It was lined with skunk cabbage; then layered with salmon innards, and split salmon heads. For two weeks, the fish heads fermented as the twice-daily high tides and low tides rinsed the pit with seawater brine. Another Tlingit treat was made by stuffing fermented salmon eggs and dried salmon strips into seal stomachs to produce an indigenous energy bar.

Although we aren't likely to ever utilize our seafood resources as completely as the first people of the Pacific rim, we can certainly improve our efficiency. For example, most fishermen fillet their catch, and discard the head and carcass. But unless they are exceptionally skilled at filleting, a substantial amount of meat remains on the carcass. This meat can be gleaned by scraping the carcass using either a sturdy spoon, or a fillet knife (which requires some facility). Gleaned scrapings can add more than ten percent to the yield from a fish. The meat can be used in a variety of recipes. Fish heads and carcasses can be used to make fish stock as well.

I hope that this "food for thought" has left you hungry for sustainable, healthful seafood. In a closure, I will leave you with a few new directions to take fish and shellfish.

Sustainable Seafood Recipes

Crab cakes

Serves 2 (double, triple, or hextuple, as necessary)

If you go crabbing in a productive area, with a large daily limit of crabs, such as here in Oregon (with a limit of 12 Dungeness crabs per person, per day), you may find yourself with a whole lot of crab meat at once, and need to figure out what to do with it. When life gives you crab meat, make crab cakes!

To cook crabs: add the crabs to a large pot of boiling water. The size of the pot will determine how many crabs you can cook at one time. Once the pot has returned to a boil, cook for 15 minutes. Using tongs, pull the cooked crabs out the pot, and rinse under cold water until cool enough to handle. Pull off the back shell, remove and discard all the innards, rinse the body, and then pick the crab meat out of the body and legs. This can be messy, so cover your counter or table with plastic and newspaper or towels to contain the mess. Use a nut cracker to crack the legs. Some of the legs can be used as picks to facilitate picking crab meat out of the legs and body. Chill the picked crab meat.

The easier path: buy crab meat.

- 1/3 cup minced onion
- 1 tsp. olive oil
- 1/2 pound crab meat
- 1 large egg white
- 1 tsp. dried parsley flakes
- 1 tsp. dried Italian seasonings
- 1/4 cup oat bran
- panko bread crumbs

In a frying pan, over medium heat, sauté onion in olive oil, until translucent. Remove from heat, and let cool. Mince or flake the crab meat and add to the sautéed onions. Add egg white, parsley, and Italian seasonings, and mix well. Add oat bran, and mix together.

Using a measuring cup, scoop the crab mixture into either 1/3- or 1/2-cup portions. Shape each portion into round, flat, crab cakes with your hands. Coat each cake with panko bread crumbs. At this point the crab cakes may be cooked, refrigerated, or frozen. They keep well in the freezer for months, if vacuum packed.

To cook: heat a little olive oil in a frying pan, over medium heat, add crab cakes, then cover. Cook 5 minutes on each side. Serve hot.

Serve with horseradish/dill sauce: 1 cup of plain yogurt or mayonnaise (or a combination of the two), 1 teaspoon of prepared horseradish, and 1 tablespoon of dried dill weed, mixed well; sweet hot mustard sauce: 3/4 cup yogurt and 1/4 cup sweet hot mustard, mixed well; cocktail sauce: 1 cup of ketchup and 1 teaspoon of prepared horseradish (or more to taste), mixed well; or another prepared sauce.

Note: A wide variety of fish cakes may be made using this recipe by substituting cooked fish meat for crab meat, including salmon, and albacore.

Crab and chanterelle couscous

Serves 2

- 1/2 of a large onion, thinly sliced, and coarsely chopped (about 1 cup)
- 2 Tbsp. olive oil
- 4 cloves garlic, thinly sliced
- 6 ounces chanterelle mushrooms, coarsely chopped
- 6 sun-dried tomatoes, in oil, drained, and diced
- 1/4 cup seafood broth
- 1/4 cup dry white wine
- freshly ground pepper to taste
- 6 ounces crab meat
- 2 Tbsp. parsley, minced
- 1/2 cup whole-wheat couscous
- 1/2 cup boiling water

In a frying pan, caramelize onion in olive oil by cooking over medium heat for about 5 minutes, then reducing heat to low. Add garlic and cook for about 10 minutes, stirring occasionally. Add mushrooms and a little water (2 Tbsp.), turn up heat to medium high and cook until the mushrooms no longer release liquid, stirring occasionally. Reduce the heat to medium. Add sun-dried tomatoes, broth and wine. Season with freshly ground pepper to taste, and mix well. Heat until the sauce is simmering, then add crab meat and stir to blend. Continue cooking to thoroughly warm the crab. Remove from heat and add parsley.

While making the sauce, in a heat-resistant bowl, pour boiling water over couscous, stir, then cover for 5 minutes. Remove cover and fluff the couscous, with a fork.

Add the crab and chanterelle sauce to the hot couscous, and mix well.

Note: Although chanterelles add a distinctive color and flavor to this dish, other varieties of mushrooms may be substituted. If seafood broth is not available, substitute vegetable broth or white wine.

Planked salmon

Serves 4 to 6

- Wild Chinook salmon fillet: 1 1/2 pounds, cut into portions

Honey mustard sauce:
- 1/2 cup Dijon mustard
- 2 Tbsp. honey
- 1/3 cup olive oil
- 1 1/2 tsp. garlic granules
- 1 Tbsp. dried Italian seasonings

In a small bowl, combine all the ingredients. Mix together until well blended.

Seasoned bread crumbs:
- 3 large cloves of garlic
- 2 Tbsp. pine nuts (also known as pignoli)
- 1 Tbsp. fennel seed, ground
- 1 tsp. dried Italian seasonings
- 1 tsp. dried parsley
- 1/2 cup oat bran
- 1/2 cup bread crumbs
- 3 Tbsp. olive oil
- 1 Tbsp. balsamic vinegar

Mince the garlic and pine nuts, and place in a small bowl. Add seasonings, oat bran, and bread crumbs. Stir until well blended. Add olive oil and balsamic vinegar, and mix until the seasoned bread crumbs are evenly moistened.

Preheat a lightly-oiled, 1 1/2 inch thick cedar plank, at 350°F for 15 minutes.

Remove plank from the oven, and place salmon skin-side-down on the plank. Slather the salmon with honey mustard sauce, coating all exposed flesh. Top with a layer of seasoned bread crumbs, about 1/4 inch thick. Bake at 350°F for 25 to 30 minutes, or until fish is done. Serve hot.

Note: Although the cedar plank enhances this dish, the recipe can also be made without a plank, by baking the fish on a lightly-oiled baking sheet at 400°F for 20 minutes. Consult guidelines that come with thin cedar planks for appropriate cooking time and temperature. Leftover honey mustard sauce can be used as the base for salad dressing. Just add some plain, non-fat yogurt, a little more olive oil, and a little water. Leftover seasoned breadcrumbs can be used to stuff artichokes or mushrooms. This recipe works for a wide variety of fish species.

Salmon en papillote (Salmon in parchment)

Serves 4 to 6

- 1 large onion, halved, then thinly sliced
- 3 Tbsp. olive oil, divided
- 1 large red pepper, cleaned, quartered, then cut into julienne strips
- 1 pound mushrooms, thinly sliced, (and coarsely chopped, if using large mushrooms)
- 1/4 cup minced garlic
- 1 1/2 pounds wild salmon fillet (preferably thin, belly meat)
- juice of half a lemon
- freshly ground pepper, to taste

In a large pan, caramelize onion in 2 Tbsp. olive oil, starting at medium heat, and reducing heat to low after about 5 minutes, stirring occasionally. Add red pepper, and continue cooking until very soft, and lightly caramelized (10 to 15 minutes). Add additional olive oil, if necessary. Remove from heat and cool to room temperature.

Steam sauté mushrooms in a large frying pan, over medium-high heat. If the mushrooms seem dry, add a small amount of water (2 Tbsp.) to the pan, to facilitate sweating the moisture out of the mushrooms. When the mushrooms appear to have given off most of their moisture, reduce heat to medium. Add 1 Tbsp. olive oil, and garlic, and then sauté until the garlic begins to brown, stirring frequently. Remove from heat and cool to room temperature. Preheat oven to 400°F.

Depending on the desired presentation, salmon may be cooked in individual-serving parchment packets, or all servings may be cooked together in one large parchment packet. For individual-serving packets: fold 4 sheets of parchment paper (16 by 24 inches) in half. Cut each folded sheet into a half heart shape, with the center of the heart along the fold. For all servings together: use rolled parchment to make a big enough packet. Open the folded, parchment hearts, and brush the inside surfaces with olive oil. Place salmon, skin side down, about an inch from the fold, on one half of the parchment heart. Season with freshly ground pepper and lemon juice. Cover with a layer of the garlic-mushroom mixture. Top with the caramelized onions and peppers. Fold the other half of the parchment heart on top of the salmon. To close each packet, double-fold the edges together, starting at the top of the heart, and continuing along the edge to the bottom end of the heart. Twist the end, or place it under the fold to seal the packet. Place packets on a baking sheet. Bake at 400°F for 20 minutes. Serve hot.

To serve, slit open packets and fold back parchment.

Note: A variety of mushrooms can be used, wild (e.g. porcini, chanterelle), or cultivated (e.g. portabella, crimini) or a combination of both. Use thin salmon fillets (less than 1 inch thick), or use scraped carcass meat. Salmon packets may be prepared ahead of time, refrigerated, then warmed to room temperature prior to baking.

Salmon loaf

Serves 6 to 8

- 1 medium sized onion, finely chopped (about 1 cup)
- 1 Tbsp. olive oil
- 1 pound cooked salmon
- 3/4 pound smoked salmon
- 1/2 cup plain non-fat yogurt
- 1 Tbsp. dried Italian seasonings
- 1 Tbsp. dried parsley
- 1/4 cup preserved lemon, rinsed then minced (optional)
- 3/4 cup oat bran

Preheat oven to 400°F. In a large frying pan, over medium heat, sauté onion in olive oil until translucent and soft, about 4 minutes. Remove from heat, and let cool. Chop or flake the salmon and smoked salmon, then add to the sautéed onions, along with the yogurt, Italian seasonings, parsley, and preserved lemon, then mix well. Add the oat bran and mix together.

Spray a loaf pan with cooking oil. Spoon the salmon mixture into the pan, and shape it into a loaf. Bake for 45 minutes in a 400°F oven. Let stand for about 5 minutes, before removing the salmon loaf from the pan and slicing. Serve hot, room temperature, or cold, with horseradish/dill sauce, sweet hot mustard sauce (see recipes under Crab cakes, above), or another prepared sauce.

Note: This recipe can be prepared ahead of time, refrigerated, then warmed to room temperature and baked just before a meal. It can also be baked ahead of time, sliced, then frozen, and thawed out just before a meal. If smoked salmon isn't available, additional cooked salmon can be used in its place. In a pinch (like on a tropical island, or in a cabin in the woods), canned Alaska salmon can be used. Additionally, other flavorful fish, such as albacore may be substituted for all of the salmon.

Penne with albacore belly meat

Serves 4-6

- 1 medium sized onion, chopped (about 1 cup)
- 8 cloves of garlic, minced (about 1/4 cup)
- 2 Tbsp. olive oil
- 1 1/2 cup fish or vegetable stock
- 1/3 cup capers, rinsed and drained
- 1/4 cup preserved lemon, rinsed, and minced (or the juice and grated zest of 1 lemon)
- 12 to 14 sun-dried tomatoes in olive oil, drained, then chopped
- 2 Tbsp. fresh parsley, minced (or 2 tsp. dried)
- 16 to 20 Greek olives, cut in thin slivers
- freshly ground pepper
- 16 ounces albacore belly meat (skinless, boneless), cut into 3/4 inch chunks
- 1/2 cup white wine
- 16 ounces dried penne pasta (whole wheat, if possible)

In a large frying pan, heat olive oil over medium heat. Add onion and garlic, and cook until limp and translucent (3 or 4 minutes), stirring frequently. Add the stock, and bring to a simmer. Gently stir in the capers, preserved lemon, sun-dried tomatoes, olives, and parsley. Season with pepper, to taste. Lay the strips of belly meat in a single layer in the simmering stock. Cook until the meat turns white on the bottom, then flip each piece (2 to 3 minutes on each side). When the meat starts breaking up, add white wine and stir to meld flavors. Cook until everything is warmed throughout. Place half of the sauce and most of the big chunks of albacore in a serving dish.

As the sauce is cooking, bring 5 to 6 quarts of water to a boil in a large pot, over high heat. Add penne pasta and cook until al dente. Drain the pasta using a colander, then add the drained pasta to the sauce, in the pan. Gently toss the penne with the sauce, cooking briefly until the sauce thoroughly coats the pasta.

Spoon the remaining sauce and albacore onto the pasta and serve, topped with toasted bread crumbs.

Note: Small albacore caught in sustainable hook-and-line fisheries are low in mercury. The belly meat of albacore is especially rich in omega-3 fish oils. If belly meat isn't available, loins may be used. This sauce may also be used with pasta of any shape. Salmon belly meat, or carcass scrapings may also be substituted for albacore. Belly meat is easiest to obtain when you purchase whole albacore directly off a fishing boat. If you have the fish butchered, ask for the belly meat as well as the loins.

Salmon head risotto

Serves 4

This recipe takes a while to make, but is a great way to utilize a part of the fish that, despite being full of protein and heart-healthy oils, is generally discarded. There is nothing awful in this often neglected odd bit of offal. Whenever a large Chinook salmon shows up in my kitchen, I make every effort to embrace the nutritive value of virtually the whole fish, by scraping the carcass and cooking the head.

In the exceedingly-slow-food style of cooking, the first step would be to catch your own fish. This could take a lot longer than you might imagine, even if you hire a fishing guide, so purchasing a fish with the head on, is a reasonable alternative. Many fishermen discard the heads of their fish. If you don't fish, consider asking a friend or acquaintance who fishes to save a fish head for you.

Once you obtain a fish, preferably at least a 10- to 12-pound salmon, fillet the fish, then remove the head and collar (the meat behind the head to where the fillet cut begins, just past the pectoral fin). Save the fillets for other meals. Depending on the size of the head and the size of your stock pot, you may need to cut the collar apart from the head. Using a sharp knife, or kitchen shears, cut out the gills, and discard them. The gills encompass several rows of rigid arches with purple, feathery filaments that enable the fish to obtain dissolved oxygen from water. Rinse the gill-free head, and collar to remove blood, then place in a large (at least 6 quart), sturdy, stock pot.

Making the stock:
To the fish head in the stock pot add:
- 8 cups water
- 1/2 onion, coarsely sliced
- 1 small carrot, or a handful of baby carrots, cut in large chunks
- 3 large cloves of garlic, crushed

- 1 tsp. whole peppercorns
- 1 tsp. dried parsley
- 1 tsp. dried Italian seasoning
- 1/2 tsp. fennel seeds
- 1 bay leaf
- 2 dried shiitake mushrooms (or a handful of dried porcini mushrooms)
- a sprig or 2 of fresh herbs (optional)

Cover the pot, and bring the stock to a boil over high heat, then turn the heat down to medium-low. Simmer for 15 minutes. Flip the head over in the stock, then remove the cheek that had been submerged in the stock and any head or collar meat that is cooked through. Set the meat aside. Cover the pot and simmer another 15 minutes. Flip the head again, and remove the second cheek, and additional head or collar meat that is cooked. Gently break apart what is left of the head, cover, and simmer for an additional 30 minutes. By now, the set-aside head meat will have cooled enough to remove and discard any bones, cartilage, and skin that was associated with it. Remove the stock pot from the heat. Cool. If you want to cool the stock quickly, set the pot in a kitchen sink half full of cold water, and stir the stock to dissipate the heat.

Using a large, fine-mesh strainer, and a slotted utensil, remove and discard all the big chunks of bone, skin, vegetables, etc., from the stock. Run the remaining stock through the strainer to remove and discard small bits that remain. Save the stock.

At this point, you can refrigerate or freeze the stock and head meat for later use. The head of a 10- to 12-pound fish will yield around 6 ounces of meat, a 25-pound fish will yield about a pound of meat. Each will yield 6 to 8 cups of stock which can be used for a variety of recipes, including soup.

Making the risotto:
- 3 cups of fish stock (divided)
- 1/2 cup wild rice
- 1/2 cup arborio rice or sushi rice
- 1 Tbsp. olive oil

- 1 medium-sized onion, chopped (about 3/4 cup)
- 4 to 6 large cloves of garlic, minced (about 2 Tbsp.)
- 1 small carrot (or 5 baby carrots), finely chopped (1/2 cup)
- 1 cup chopped mushrooms (wild or domestic)
- 2 tsp. dried parsley
- 2 tsp. dried Italian seasoning
- freshly ground pepper, to taste
- 6 to 8 sun-dried tomatoes in olive oil, drained, then chopped
- 8 to 12 ounces fish-head meat

Place 2 cups of fish stock and 1 cup of water in a small sauce pan, and bring to a boil. Add the wild rice, stir, cover, and turn the heat down to medium low. Set a timer for 50 minutes.

While the wild rice is cooking, chop the onion, garlic, carrot, and mushrooms. In a large frying pan, over medium heat, add olive oil, then sauté onion, garlic, and carrot until soft, about 8 minutes, stirring frequently. Add mushrooms and about a tablespoon of water and cook until the mushrooms have given off all their moisture, stirring occasionally. Add parsley, Italian seasoning, pepper, and arborio (or sushi) rice. Mix well and cook for 1 to 2 minutes. When the timer from the wild rice reads "20 minutes," pour the wild rice and its stock into the large pan, stir, and cover. Continue to cook on medium heat, stirring occasionally. Warm the remaining stock in the small sauce pan that the wild rice was just in.

Coarsely chop the fish-head meat. When the timer reads "10 minutes," add the fish-head meat and the sun-dried tomatoes to the risotto, stir well, and add the remaining fish stock. Cover and continue cooking, stirring occasionally.

Towards the last few minutes of cooking, determine how soupy the risotto is, and how wet or dry you want it. Slightly soupy risotto has a nice consistency. To dry out the risotto, cook with the cover ajar (or off entirely). If the risotto is too dry, add a little water.

Serve hot, topped with toasted breadcrumbs.

Note: This recipe can also be made with a couple of smaller fish heads or with a smaller fish head and the fish carcass. Other varieties of fish heads may also be used. Fish teeth can be very sharp so keep your fingers away from their teeth. One tool that you might find useful is a fish-handling, knife-blade-blocking glove.

Smoked oyster risotto

Serves 2 as a main dish, 4 as a side dish

- 2 tsp. olive oil
- 1 small onion, chopped (1/2 cup)
- 3 to 4 large cloves garlic, minced (3 Tbsp.)
- 4 large mushrooms, chopped (1/2 cup)
- 1 small carrot (or 5 baby carrots), grated or minced (1/2 cup)
- 1/2 cup arborio rice or sushi rice
- 1 scant tsp. Italian seasonings
- 1 scant tsp. dried parsley
- freshly grated pepper, to taste
- 1 1/2 cups vegetable or seafood broth
- 1 1/2 cups water (more as needed)
- 6 ounces smoked oysters (about 6 medium size), chopped
- 6 to 8 sun-dried tomatoes in olive oil, drained, then chopped

In a large frying pan, heat olive oil over medium heat. Add onion and garlic and sauté until nearly translucent. Add mushrooms and carrots, and cook until tender, stirring occasionally. Add rice, Italian seasonings, parsley, and pepper. Stir well, and cook for 1 to 2 minutes. Add broth, stir, and cover. Set timer for 20 minutes. Stir every few minutes, checking to make sure rice is not sticking to the pan. When 10 minutes remain on the timer, add oysters and sun-dried tomatoes. Stir well, then add about a cup of water. Cook for 10 more minutes, stirring occasionally, adding more water as necessary.

Towards the last few minutes of cooking, determine how soupy the risotto is, and how wet or dry you want it. Slightly soupy risotto has a nice consistency. To dry out the risotto, cook with the cover ajar (or off entirely). If the risotto is too dry, add a little water.

Serve, topped with toasted bread crumbs.

Note: This recipe can also be made using smoked mussels.

Smoked albacore pizza

Serves 4

- Pizza dough for 1 pizza, store-bought or homemade (recipe below)

Make the smoked albacore sauce:
- 1 Tbsp. olive oil
- 3/4 cup minced onion
- 3 to 4 cloves garlic, minced
- 1 can (28-oz.) crushed tomatoes with basil
- 1 Tbsp. dried basil
- 1/2 tsp. dried oregano
- 2 Tbsp. chipotle sauce
- 2 cups smoked albacore, chopped

In a large frying pan, heat olive oil over medium heat. Sauté onion and garlic, until transparent. Add tomatoes, seasonings, and chipotle sauce, and mix well. Cook until the sauce is reduced to a very thick texture, stirring occasionally. Add the smoked albacore. Stir well to meld flavors. Remove from heat, and let cool. Preheat oven to 400°F. If using a pizza stone, preheat the stone.

Prepare mushrooms:
- 1/2 pound of mushrooms, sliced (wild, cultivated, or a mixture)

Place mushrooms in a large pan with a little water (2 to 3 Tbsp.). Steam sauté over medium-high heat, 3 or 4 minutes, until mushrooms release their liquid. Remove from heat, drain, and let cool.

Assemble the pizza:
On a lightly floured surface, roll out the dough for 1 pizza until 1/8- to 1/4-inch thick, shaping it to fit a pizza stone or pan. Dust a pizza stone with cornmeal, or lightly oil a pizza pan. Lift

the dough onto the pizza stone/pan and turn up the edges to form a rim. Paint the albacore sauce, evenly, inside of the rim, then top the sauce with the mushrooms. Bake at 400°F for 20 to 25 minutes.

Homemade pizza dough (for 2 pizzas):
- 1 cup very warm water
- 1 Tbsp. (or package) rapid-rise yeast
- 1 1/2 cups semolina flour
- 1 1/2 cups whole wheat flour
- 2 tsp. sugar
- 1 tsp. salt
- 1/2 tsp. garlic granules
- 2 Tbsp. fresh rosemary, minced
- 2 Tbsp. olive oil

Dissolve the yeast in warm water, and let proof for 5 minutes. Mix the dry ingredients and rosemary together in a bowl. Add the olive oil to the dissolved yeast, then add all the liquid ingredients to the flour mixture, stirring well to mix. Turn the dough onto a lightly floured surface and knead until smooth (or knead with a bread hook or food processor). Form the dough into a ball and place it in a large bowl that has been lightly oiled with olive oil. Roll the dough ball in the bowl to coat with oil. Cover with a damp cloth and let rise in a warm, draft-free space, until double (about 30 minutes for rapid-rise yeast). Punch down dough, re-form into a ball, cover and let rise for a second time (30 more minutes). Punch down, divide in half, roll out half the dough and save the other half for another pizza. Dough can be flattened into a disk and frozen at this point.

Note: Assembling the pizza on a hot pizza stone will pre-cook the dough a bit, and insure that the dough will not be soggy. The smoked albacore sauce can also be used on pasta or lasagna. Smoked herring could be substituted for the smoked albacore. Minced chipotle chilies in adobo sauce, 1 tsp. powdered chipotle chili, or another smoky, spicy hot sauce may be substituted for the chipotle sauce.

Smoked salmon enchiladas

Serves 3 to 4 as a main dish, 6 to 8 as a side dish

- 1 large onion, thinly sliced and coarsely chopped (about 2 cups)
- 1 large red pepper, cut into julienne strips, then coarsely chopped
- 2 Tbsp. olive oil
- 1 pound of mushrooms, thinly sliced, then coarsely chopped
- 2 cups smoked salmon, flaked or chopped
- 1 tsp. ground cumin
- 6 to 8 whole wheat tortillas (8 inch size)
- 1 14-oz. can mild enchilada sauce (divided)
- 2 Tbsp. chipotle sauce (see note)

In a large pan, caramelize onion and red pepper in olive oil: cook over medium heat for about 5 minutes, then reduce heat for about 10 minutes, stirring frequently. Place mushrooms and a little water (2 to 3 Tbsp.) in a large frying pan and steam sauté over medium-high heat, until mushrooms no longer release liquid. Let cool. Add caramelized onions and peppers, smoked salmon, and cumin, to the mushrooms, mixing to blend. Preheat oven to 375°F.

Spray a 9 x 14 inch baking dish with canola oil. Pour half of the enchilada sauce into the baking dish. Dip each tortilla in the sauce in the baking dish, to coat. Divide the filling into 6 to 8 parts (equal to number of tortillas). Spoon the filling across the center of each tortilla, fold over one side, and continue to roll it up around the filling. Place the filled enchiladas in the baking dish, loose end down.

Add chipotle sauce to the remaining enchilada sauce, and mix to blend. Pour the sauce over the rolled enchiladas, in the baking dish. Cover with aluminum foil, and bake at 375°F for 35 minutes. Remove foil and bake for an additional 5 minutes.

Note: Minced chipotle chilies in adobo sauce, 1 tsp. powdered chipotle chili, or another smoky, spicy hot sauce may be substituted for chipotle sauce.

Mussels fra diavolo

Serves 2

- 2 dozen mussels in their shells (or 8 to 10 ounces of precooked mussel meats and 1 cup of mussel broth, recipe below)
- 1 Tbsp. olive oil
- 1 small or 1/2 medium-sized sweet onion, chopped
- 4 to 6 garlic cloves, minced
- 1 can (28-oz.) crushed tomatoes with basil
- 1/2 cup coarsely chopped roasted red pepper
- 2 Tbsp. dried basil
- 1 tsp. dried oregano
- 1 tsp. hot red pepper flakes
- dash of cayenne pepper
- 6 ounces linguine

Scrub the mussels and remove their beards.

In a large frying pan, heat olive oil over medium heat. Add the garlic and onion, and cook until limp. Add tomatoes, red pepper, dried basil, oregano, pepper flakes, and cayenne. Simmer for about 20 minutes, stirring frequently, until sauce is thick. Add the mussels in their shells. Cover and cook for about 10 minutes, until all shells are open, and their nectar has been released into the sauce, stirring occasionally.

As the sauce is cooking, bring 5 to 6 quarts of water to a boil in a large pot, over high heat. Add the linguine and cook until al dente. Drain the pasta using a colander, then add about half the sauce to the drained pasta in a large bowl. Gently toss the linguine with the sauce, until the sauce thoroughly coats the pasta.

Serve the linguine topped with the remaining sauce and the mussels.

Pre-cooking mussels:

Steam up to 2 dozen mussels over 3 to 4 cups of water until all the mussels have opened, about 5 minutes, remove mussels from the pot and rinse with cold water to cool. Save the broth. The same broth can be used to steam several batches of mussels. When the mussels are cool enough to handle, pull the meat out of the shells and remove the beards. Rinse to remove any sand. Drain well, then vacuum pack, or place in a freezer container, and pour strained broth over mussels, then freeze for later use. When ready to use: warm the mussels in their broth until all the broth has melted. Remove mussels from broth and set aside. To use in this recipe: add 1 cup of broth to the seasoned tomato/red pepper sauce and simmer until the sauce is very thick, about 20 minutes. Add mussel meats and cook for an additional 5 to 6 minutes. Stir well and serve over pasta.

Spicy linguine with mussels in a red pepper and bruschetta sauce

Serves 2 as a main dish, 4 as a side dish

- 2 Tbsp. olive oil
- 1 red pepper, diced
- 6 cloves garlic, thinly sliced
- 1 tsp. hot red pepper flakes
- 2/3 cup bruschetta sauce
- 1/3 cup white wine
- 2 Tbsp. minced fresh parsley
- freshly ground pepper to taste
- 2 dozen mussels in their shells (or 8 to 10 ounces pre-cooked mussel meats and 1/2 cup mussel broth, see recipe above)
- 6 ounces linguine

In a large frying pan, heat olive oil, over medium heat. Add red pepper, garlic, and hot pepper flakes and sauté for 5 to 6 minutes, or until soft, stirring occasionally. Add bruschetta sauce, wine, parsley, and freshly ground pepper. Stir to mix and simmer for a few minutes. Then add the mussels and cover. Cook until all the mussels have opened and released their nectar. If using mussel meats and broth, cook for 3 to 5 minutes. Remove the mussels to a serving bowl.

While the sauce is cooking, bring a large pot of water to a boil and cook the linguine until it is al dente. Drain the linguine, and add it to the sauce, cooking for several minutes until the pasta absorbs the remaining liquid of the sauce.

Serve topped with the mussels.

Note: If bruschetta sauce is not available, feel free to substitute 1/2 cup chopped tomatoes, 1/2 cup minced onion, 2 Tbsp. minced parsley, and 2 Tbsp. minced basil.

Pesto pink shrimp on spaghetti

Serves 2 as a main dish, 4 as a side dish

- 2 tsp. olive oil
- 1/3 cup onion, thinly sliced and coarsely chopped
- 1 cup of mushrooms, thinly sliced and coarsely chopped
- 8 sun-dried tomatoes, in olive oil, drained, then chopped
- 1/4 to 1/3 cup pesto sauce (see recipe below)
- 1/2 cup white wine
- 1/2 cup broth (seafood or vegetable broth, or water)
- freshly ground pepper, to taste
- 8 ounces cooked Oregon pink shrimp meat (cocktail shrimp)
- 6 ounces spaghetti

In a large frying pan, heat olive oil, over medium heat. Add onions and mushrooms and sauté until the mushrooms have given off their moisture, and the onions are translucent. Add the sun-dried tomatoes, pesto, wine, stock, and pepper, stirring to blend well. Cook for several minutes to meld flavors. If the sauce thickens too much, add a little water from the cooking pasta. Gently stir the cooked shrimp meat into the pesto mixture, cooking only long enough to heat throughout. Serve immediately.

While the sauce is cooking, bring 5 to 6 quarts of water to a boil in a large pot, over high heat. Add the spaghetti and cook until al dente. Drain using a colander, then in a large bowl, add about half of the sauce to the drained spaghetti. Gently toss the pasta with the sauce.

Spoon the remaining sauce onto the spaghetti, and serve topped with grated Parmesan cheese.

Pesto sauce (without cheese):
- 8 large cloves garlic
- 1 clove elephant garlic, roasted (optional)
- 1/4 cup pine nuts (pignoli)
- 1/4 pound of basil leaves (about 1 1/2 to 2 large bunches, stems removed)

- 1/4 cup fresh parsley, stems removed
- 1/2 cup olive oil

Wash and towel-dry the basil leaves and parsley. In a food processor or blender, finely mince the garlic and pine nuts. Add the basil to the minced garlic, tearing any large leaves into smaller pieces. Pulse to mince and blend the basil into the garlic mixture. Add the parsley to the food processor, and pulse to blend. After the basil and parsley are uniformly minced, add the olive oil and pulse to incorporate all the ingredients.

At this point, you can use or freeze the pesto. If you are storing it in the refrigerator, cover the surface with a thin layer of olive oil, to prevent darkening.

Note: Crab meat or smoked salmon can be substituted for the shrimp meat.

End notes

1. Quadfasel, D. 2005. The Atlantic heat conveyor slows. Nature 438:565-566.

2. Bryden, H.L., H.R. Longworth, and S.A. Cunningham. 2005. Slowing of the Atlantic meridional overturning circulation at 25°N. Nature 438:655-657.

3. Schiermeier, Q. 2007. Ocean circulation noisy, not stalling. Nature 448:844-45.

4. Kurlansky, M. 1997. Cod: A biography of the fish that changed the world. Walker Publishing Company Inc. New York.

5. Simon, A.W. 1984. Neptune's revenge: The ocean of tomorrow. Franklin Watts. New York.

6. Diaz, R.J. and R. Rosenberg. 2008. Spreading dead zones and consequences for marine ecosystems. Science 321(5891):926-929.

7. NOAA-supported scientists find changes to Gulf of Mexico dead zone. August 9, 2010. NOAA News. Sources: http://www.noaanews.noaa.gov/stories2010/20100809_deadzone.html and http://www.gulfhypoxia.net/research/Shelfwide%20Cruises/2010/PressRelease2010.pdf. Accessed on 5-24-2012.

8. Gewin, V. 2010. Dead in the water. Nature 466:814-816.

9. Mitchell, A. 2009. Seasick: Ocean change and the extinction of life on Earth. University of Chicago Press. Chicago.

10. Barton, A., B. Hales, G.G. Waldbusser, C. Langdon and R.A. Feely. 2012. The Pacific oyster, *Crassostrea gigas*, shows negative correlation to naturally elevated carbon dioxide levels: Implications for near-term ocean acidification effects. Limnology and Oceanography 57:698-710.

11. Worm, B., E.B. Barbier, N. Beaumont, J.E. Duffy, C. Folke, B.S. Halpern, J.B.C. Jackson, H.K Lotze, F. Micheli, S.R. Palumbi, E. Sala, K.A. Selkoe, J.J. Stachowicz, and R. Watson. 2006. Impacts of biodiversity loss on ocean ecosystem services. Science 314(5800):787-790.

12. Nettleton, J.A. 1987. Seafood and health. Osprey books. Huntington, NY.

13. Simopoulos, A.P. and J. Robinson. 1999. The omega diet: the lifesaving nutritional program based on the diet of the Island of Crete. HarperPerrenial, New York.

14. Lichtenstein, A.H., L.J. Appel, M. Brands, M. Carnethon, S. Daniels, H.A. Franch, B. Franklin, P. Kris-Etherton, W.S. Harris, B. Howard, N. Karanja, M. Lefevre, L. Rudel, F. Sacks, L. Van Horn, M. Winston, and J. Wylie-Rosett. 2006. Summary of American Heart Association Diet and Lifestyle Recommendations Revision 2006. Arteriosclerosis, Thrombosis, and Vascular Biology. 26:2186-2191.

15. Krümmel, E.M., R.W. Macdonald, L.E. Kimpe, I. Gregory-Eaves, M.J. Demers, J.P. Smol, B. Finney, and J.M. Blais. 2003. Delivery of pollutants by spawning salmon: fish dump toxic industrial compounds in Alaskan lakes on their return from the ocean. Nature 425:256.

16. Blais, J.M., L.E. Kimpe, D. McMahon, B.E. Keatley, M.L. Mallory, M.S.V. Douglas, and J.P. Smol. 2005. Arctic seabirds transport marine-derived contaminants. Science 309(5733):445.

17. EPA Water: Outreach and Communication. What You Need to Know about Mercury in Fish and Shellfish. 2004. Source: http://www.epa.gov/waterscience/fishadvice/advice.html. Accessed on 5-18-2012.

18. EPA Fact Sheet: National listing of fish advisories. August 2004. Source: http://nepis.epa.gov/Exe/ZyPURL.cgi?Dockey=6000031J.TXT. Accessed on 5-26-2012.

19. EPA Water: Methyl mercury. Human health criteria: Methylmercury fish tissue criterion. 2010. Source: http://www.epa.gov/waterscience/criteria/methylmercury/merctitl.pdf. Accessed on 5-18-2012.

20. Peterson, S.A., J. Van Sickle, A.T. Herlihy, and R.M. Hughes. 2007. Mercury concentration in fish from streams and rivers throughout the western United States. Environmental Science and Technology 41:58-65.

21. Pollution overtaking lakes, rivers. August 24, 2004. Sources: http://www.cbsnews.com/stories/2004/08/24/tech/main638130.shtml and EPA Fact Sheet: National listing of fish advisories. August 2004. Source: http://nepis.epa.gov/Exe/ZyPURL.cgi?Dockey=6000031J.TXT. Accessed on 5-18-2012.

22. FDA and EPA. 2004. What you need to know about mercury in fish and shellfish. March 2004. Source: http://www.fda.gov/food/resourcesforyou/consumers/ucm110591.htm. Accessed on 6-6-2013.

23. Hibbeln, J.R., J.M. Davis, C. Steer, P. Emmett, I. Rogers, C. Williams, and J. Golding. 2007. Maternal seafood consumption in pregnancy and

neurodevelomental outcomes in childhood (ALSPAC study): an observational cohort study. Lancet 369:578-85.

24. East Lake; Newberry Crater Fish Advisory: Fish mercury advisory revised for East Lake, Newberry Crater, Deschutes County. July 9, 1996. Source: http://public.health.oregon.gov/HealthyEnvironments/Recreation/Pages/eastlk.aspx. Accessed on 5-18-2012.

25. Moestrup, Ø., R. Akselman, G. Cronberg, M. Elbraechter, S. Fraga, Y. Halim, G. Hansen, M. Hoppenrath, J. Larsen, N. Lundholm, L.N. Nguyen, and A. Zingone (eds.). (2009 onwards). IOC-UNESCO Taxonomic Reference List of Harmful Micro Algae. Available online at http://www.marinespecies.org/HAB. Accessed on 5-15-2012.

26. King James Bible. Source: http://kingjbible.com/exodus/7.htm. Accessed on 5-18-2012.

27. Jewish Publication Society Bible. Source: http://www.jewishvirtual library.org/jsource/Bible/Deuter14.html. Accessed on 5-18-2012.

28. Jewish Publication Society Bible. Source: http://www.jewishvirtual library.org/jsource/Bible/Leviticus11.html. Accessed on 5-18-2012.

29. pp. 285-286 In: Vancouver, G. 1798. A voyage of discovery to the north Pacific Ocean and around the world. Vol 2. G.G. and J. Robinson, London.

30. Gessner, B.D. and J.P. Middaugh. 1995. Paralytic Shellfish Poisoning in Alaska: A 20-Year Retrospective Analysis. American Journal of Epidemiology 141:766–70.

31. Sharifzadeh K., N. Ridley, R. Waskiewicz, P. Luongo, G.F. Grady, A. DeMaria, R.J. Timperi, J. Nassif, M. Sugita, V. Gehrman, P. Peterson, A. Alexander, R. Barrett, K. Ballentine, J.P. Middaugh and I. Somerset. 1991. Epidemiologic notes and reports Paralytic Shellfish Poisoning -- Massachusetts and Alaska, 1990. Morbidity and Mortality Weekly Reports 40(10):157-161.

32. McLaughlin, J.B., D.A. Fearey, T.A. Esposito, and K.A. Porter. 2011. Paralytic Shellfish Poisoning — Southeast Alaska, May–June 2011. Morbidity and Mortality Weekly Reports 60(45):1554-1556.

33. Watkins, S.M., A. Reich, L.E. Fleming, and R. Hammond. 2008. Neurotoxic Shellfish Poisoning. Marine Drugs 6(3):431–455. Published online 7-12-2008.

34. Quilliam, M.A. and Wright, J.L.C. 1989. The amnesic shellfish poisoning mystery. Analytical Chemistry 61:1053A-1059A. Cited in: Sullivan, J.J. 1993. Methods of analysis for algal toxins: dinoflagellate and

diatom toxins. pp 29-48 *In*: Algal toxins in seafood and drinking water. I.R. Falconer (ed.). Academic Press, London.

35. Halstead, B.W. 1965. In: Poisonous and venomous marine animals of the world. Darwin. Princeton, N.J., Vol. 1. Cited in: Ragelis E.P. 1984. Ciguatera seafood poisoning: overview. Pp. 25-36 *In*: Seafood toxins. E.P. Ragelis (ed.). American Chemical Society. Washington, D.C.

36. Ragelis, E.P. 1984. Ciguatera seafood poisoning: overview. Pp. 25-36 *In*: Seafood toxins. E.P. Ragelis (ed.). American Chemical Society. Washington, D.C.

37. Department of Health and Human Services, Centers for Disease Control and Prevention, National Center for Environmental Health. Harmful algal blooms: Ciguatera fish poisoning. Sources: http://www.cdc.gov/nceh/ciguatera/default.htm and http://www.cdc.gov/nceh/ciguatera/fish.htm. Accessed on 5-18-2012.

38. FDA, Center for Food Safety and Applied Nutrition, Office of Food Safety. 2011. Chapter 6 Natural toxins. Pp. 99-112 *In*: Fish and Fishery Products Hazards and Controls Guidance, Fourth Edition. Source: http://www.fda.gov/downloads/Food/GuidanceComplianceRegulatoryInformation/GuidanceDocuments/Seafood/UCM251970.pdf. Accessed on 5-21-2012.

39. World Health Organization. 2007. International travel and health: situation as on 1 January 2007. Source: http://whqlibdoc.who.int/publications/2007/9789241580397_eng.pdf. Accessed on 5-26-2012.

40. Fleming, L.E., D. Katz, J.A. Bean, and R. Hammond. 2001. Epidemiology of seafood poisoning. Pp. 287-310 *In*: Foodborne disease handbook, 2nd edition Vol. 4: Seafood and environmental toxins. Y.H. Hui, D. Kitts, and P.S. Stanfield (eds.). Marcel Dekker, Inc. New York.

41. Facts and Details: Fugu (blowfish) in Japan. 2009. Source: http://factsanddetails.com/japan.php?itemid=649&catid=19&subcatid=123 . Accessed on 5-26-2012. and Lyon, N. Travel intelligence: Japans Kamikase Kuisine. Source: http://www.travelintelligence.com/travel-writing/japan%C2%92s-kamikaze-kuisine. Accessed on 5-27-2012.

42. Davis, W. 1988. Zombification. Science 240(4860):1715-1716.

43. Yentsch, C.M. 1984. Paralytic shellfish poisoning: an emerging perspective. Pp. 9-23 *In*: Seafood toxins. E.P. Ragelis (ed.). American Chemical Society. Washington, D.C.

44. Altman, L.K. 1993. What's sauce for the oyster may also keep the doctor away. The New York Times. Health. Published: Oct. 19,1993.

Source: http://query.nytimes.com/gst/fullpage.html?sec=health&res=
9F0CEEDA1531F93AA25753C1A965958260. Accessed on 5-15-2013.

45. Sun, Y. and J.D. Oliver. 1995. Hot sauce: no elimination of *Vibrio vulnificus* in oysters. Journal of Food Protection 58(4):441-442.

46. FDA *Vibrio vulnificus* health education kit. *Vibrio vulnificus* fact sheet. Source: http://www.fda.gov/Food/ResourcesForYou/HealthEducators/ucm085365.htm. Accessed on 5-21-2012.

47. Mirza, R.A., J.-J. Poisson, G. Fisher, A. D'Aniello, P. Spinelli, and G. Ferrandino. 2005. Do marine mollusks possess aphrodisiacal properties? Presented at The 229th American Chemical Society National Meeting, in San Diego, CA, March 13-17, 2005. Abstract only.

48. Marean, C.W., M. Bar-Matthews, J. Bernatchez, E. Fisher, P. Goldberg, A.I.R. Herries, Z. Jacobs, A. Jerardino, P. Karkanas, T. Minichillo, P.J. Nilssen, E.Thompson, I. Watts, and H.M. Williams. 2007. Early human use of marine resources and pigment in South Africa during the Middle Pleistocene. Nature 449:905-908.

49. Yellen, J.E., A.S. Brooks, E. Cornelissen, M.J. Mehiman, and K. Stewart. 1995. A Middle Stone Age worked bone industry from Katanda, upper Semliki Valley, Zaire. Science 268(5210):553-556.

50. O'Connor, S., R. Ono, and C. Clarkson. 2011. Pelagic fishing at 42,000 years before the present and the maritime skills of modern humans. Science 334 (6059):1117-1121.

51. Borgerson, S.G. 2009. The National Interest and the Law of the Sea. Council on Foreign Relations Council Special Report No. 46 May 2009.

52. Kura, Y.C. Revenga, E. Hoshino, and G. Mock. 2005. Fishing for answers: making sense of the global fish crisis. World Resources Institute. Washington, D.C.

53. Franklin, H.B. 2007. The most important fish in the sea: menhaden and America. Island Press. Washington, D.C.

54.Tilapia history and tilapia future. 2008. Source: http://www.aquaticcommunity.com/tilapia/history.php. Accessed on 6-24-2012.

55. Monterey Bay Aquarium Seafood Watch. Source: http://www.montereybayaquarium.org/cr/cr_seafoodwatch/download.aspx. Accessed on 7-12-2012.

56. Blue Ocean Institute Seafood Guide. Source: http://blueocean.org/seafood/. Accessed on 5-25-2012.

57. Seafood Selector: fish choices that are good for you and the ocean. Source: http://www.edf.org/sites/default/files/1980_pocket_seafood_selector.pdf. Accessed on 7-11-2012.

58. Sushi Selector: sushi choices that are good for you and the ocean. Source: http://apps.edf.org/documents/8683_sushi_pocket.pdf. Accessed on 7-12-2012.

59. Lee, M., (ed.). 2000. Seafood Lover's Almanac. Living Oceans Program, National Audubon Society. Islip, NY.

60. Baldwin, C.C. And J.H. Mounts. 2003. One fish, two fish, crawfish, bluefish: the Smithsonian sustainable seafood cookbook. Smithsonian Books. Washington, D.C.

61. Clover, C. 2006. The end of the line: how overfishing is changing the world and what we eat. University of California Press, Berkeley.

62. Grescoe, T. 2008. Bottomfeeder: how to eat ethically in a world of vanishing seafood. Bloomsbury. New York.

63. Trenor, C. 2009. Sushi: a guide to saving the oceans one bite at a time. North Atlantic Books, Berkeley, California.

64. Greenberg, P. 2010. Four fish: the future of the last wild food. Penguin Press. New York.

65. Navigating sustainability: how to make the best seafood choices at Whole Foods Market. Source: http://wholefoodsmarket.com/seafood-ratings/index.php. Accessed on 7-14-2012.

66. Roberts, C. 2007. The Unnatural History of the Sea. Island Press. Washington, D.C.

67. Francis, R.I.C.C. and M.R. Clark. 2005. Sustainability issues for orange roughy fisheries. Bulletin of Marine Science 76:337-351.

68. Should we eat the orange roughy? Mar-Eco: patterns and processes of the ecosystems of the northern mid-Atlantic. Source: http://www.mar-eco.no/learning-zone/backgrounders/deepsea_life_forms/orange_roughy_story. Accessed on 7-17-2012.

69. Knecht, G.B. 2006. Hooked: pirates, poaching, and the perfect fish. Rodale. Emmaus, PA.

70. Camhi, M.D., E.K. Pikitch, and E.A. Babcock. 2008. Sharks of the open ocean: biology, fisheries and conservation. Blackwell Publishing. Oxford.

71. Cailliet, G.M., A.H. Andrews, E.J Burton, D.L Watters, D.E. Kline, and L.A. Ferry-Graham. 2001. Age determination and validation studies of

marine fishes: do deep-dwellers live longer? Experimental Gerontology 36:739-764.

72. Camhi, M.D., S.V. Valenti, S.V. Fordham, S.L. Fowler, and C. Gibson. 2009. The Conservation Status of Pelagic Sharks and Rays. Report of the IUCN Shark Specialist Group, Pelagic Shark Red List Workshop. IUCN Species Survival Commission's Shark Specialist Group. Newbury, UK.

73. Kurtenbach, E. 2011. Yao Ming, Richard Branson team up to stop shark fin trade. Christian Science Monitor. Source: http://www.csmonitor.com/World/Latest-News-Wires/2011/0923/Yao-Ming-Richard-Branson-team-up-to-stop-shark-fin-trade. Accessed on 9-29-2012.

74. Clarke, S.C., M.K. McAllister, E.J. Milner-Gulland, G.P. Kirkwood, C.G.J. Michielsens, D.J. Agnew, E.K. Pikitch, H. Nakano, and M.S. Shivji. 2006. Global estimates of shark catches using trade records from commercial markets. Ecology Letters 9: 1115–1126.

75. Vincent, A.C.J., S.J. Foster, and H.J. Koldewey. 2011. Conservation and management of seahorses and other Syngnathidae. Journal of Fish Biology 78:1681–1724.

76. Cohen, T. 2012. Seahorses 'are facing oblivion in 10 years' after stocks are savaged by Chinese medicine industry. Mail Online. Science & Tech. 3 August 2012. Source: http://www.dailymail.co.uk/sciencetech/article-2183501/Seahorses-facing-oblivion-10-years-stocks-savaged-Chinese-medicine-industry.html#ixzz23RUGkQ5B. Accessed on 8-13-2012.

77. Scales, H. 2009. Poseidon's steed: The story of seahorses, from myth to reality. Gotham Books, Penguin Group. New York.

78. Project seahorse: seahorse conservation and assessment. Source: http://seahorse.fisheries.ubc.ca/what-we-do/save-seahorses/seahorse-conservation-assessment. Accessed on 7-21-2012.

79. Ellis, R. 2008. Tuna: love, death, and mercury. Alfred A. Knopf. New York.

80. Watson, L. 2012. Bluefin tuna sold for a record $736,000 (that's $96 per slice of sushi) in Japan – and the price will only get higher as stocks run dry. Mail Online. 6 January 2012. Source: http://www.dailymail.co.uk/news/article-2082666/Giant-bluefin-tuna-sold-record-472-000-Japan.html. Accessed on 8-12-2012.

81. Giant Bluefin Tuna Fetches Record $396,000 In Tokyo Auction. Huffington Post.1-05-2011. Source: http://www.huffingtonpost.com/2011/01/05/bluefin-tuna-record-tokyo-auction_n_804553.html. Accessed on 8-20-2012.

82. Risenhoover, A.D. 2012. Atlantic Highly Migratory Species; 2012 Atlantic Bluefin Tuna Quota Specifications. Federal Register. March 16, 2012. Source: https://www.federalregister.gov/articles/2012/03/16/2012-6453/atlantic-highly-migratory-species-2012-atlantic-bluefin-tuna-quota-specifications. Accessed on 8-16-2012.

83. Safina, C. 1997. Song for the blue ocean: encounters along the world's coasts and beneath the seas. Henry Holt. New York.

84. Safina, C. 2011. A sea in flames: the Deepwater Horizon oil blowout. Crown Publishers. New York.

85. IUCN 2012. IUCN Red List of Threatened Species. Version 2012.1. Source: www.iucnredlist.org. Accessed on 8-16-2012.

86. Musick, J.A. and J.K. Ellis. 2004. Constraints on sustainable marine fisheries in the United States: a look at the record. American Fisheries Society Symposium 43:45-66.

87. FAO Topic Fact Sheets. Assessing fishing capacity and over-capacity. FAO Fisheries and Aquaculture Department. 27 May 2005. Source: www.fao.org/fishery/topic/14858/en. Accessed on 9-17-2012.

88. Sumaila, U.R., A.S. Khan, A.J. Dyck, R. Watson, G. Munro, P. Tydemers, and D. Pauly. 2010. A bottom-up re-estimation of global fisheries subsidies. Journal of Bioeconomics 12:201-225.

89. Kura et al., 2005.

90. Hall et al., M.A., D.L. Alverson, and K.I. Metuzals. 2000. By-catch: problems and solutions. Marine Pollution Bulletin Vol. 41:204-219.

91. Hall et al., 2000.

92. Brothers, N.P., J. Cooper, and S. Lokkeborg. 1999. Incidental catch of seabirds by longline fisheries: worldwide review and technical guidelines for mitigation. FAO Fisheries Circular No. 937. FAO. Rome.

93. Anderson, O.R.J., C.J. Small, J.P. Croxall, E.K. Dunn, B.J. Sullivan, O. Yates, and A. Black. 2011. Global seabird bycatch in longline fisheries. Endangered Species Research 14: 91–106. Source: http://www.int-res.com/articles/esr_oa/n014p091.pdf. Accessed on 8-24-2012.

94. Dietrich, K.S., J.K. Parrish, and E.F. Melvin. 2009. Understanding and addressing seabird bycatch in Alaska demersal longline fisheries. Biological Conservation 142(11):2642-2656.

95. Seabird Bycatch Mitigation Factsheets. BirdLife International global seabird programme homepage. Source: http://www.birdlife.org/seabirds/bycatch/albatross.html. Accessed on 10-7-2012.

96. Watson, J.W., S.P. Epperly, A.K. Shah, and D.G. Foster. 2005. Fishing methods to reduce sea turtle mortality associated with pelagic longlines. Canadian Journal of Fisheries and Aquatic Sciences 62:965-981.

97. Harrington, J.M., R.A. Meyers, and A.A. Rosenberg. 2005. Wasted fishery resources: discarded by-catch in the USA. Fish and Fisheries 6:350-361.

98. Gallaway, B.J., M. Longnecker, J.G. Cole, and R.M. Meyer. 1998. Estimates of shrimp trawl bycatch of red snapper (Lutjanus campechanus) in the Gulf of Mexico. Pp. 817-839 *In*: Fishery stock assessment models : proceedings of the International Symposium on Fishery Stock Assessment Models for the 21st Century, October 8-11, 1997, Anchorage, Alaska. F. Funk, T.J. Quinn, J. Heifetz, J.N. Ianelli, J.E. Powers, J.F. Schweigert, P.J. Sullivan, and C.I. Zhang (eds.). Alaska Sea Grant College Program, Report AK-SG-98-01. Fairbanks, AK.

99. FAO. 2002. Fishing Technology Equipments. Turtle Excluder Device (TED). Technology Fact Sheets. Text by V. Crespi and J. Prado. In: FAO Fisheries and Aquaculture Department website. Source: http://www.fao.org/fishery/equipment/ted/en. Accessed on 10-8-2012.

100. Macfadyen, G., T. Huntington, and R. Cappell. 2009. Abandoned, lost or otherwise discarded fishing gear. FAO Fisheries and Aquaculture Technical Paper No. 523.

101. Valdemarsen, J.W. and P. Suuronen. 2003. Modifying fishing gear to achieve ecosystem objectives. Pp. 321-341 *In*: M. Sinclair and G. Valdimarsson (eds.). Responsible fisheries in the marine ecosystem. FAO and CABI Publishing. Wallingford, UK.

102. Norwegian fisheries management, our approach of discard of fish. Norwegian Ministry of Fisheries and Coastal Affairs. Source: http://www.regjeringen.no/upload/FKD/Brosjyrer%20og%20veiledninger/fact_sheet_discard.pdf. Accessed on 8-29-2012.

103. Wigan, M. 1998. The last of the hunter gatherers: fisheries crisis at sea. Swan-Hill Press. Shrewsbury, England.

104. Daskalov, G.M., A.N. Grishin, S. Rodionov, and V. Mihneva. 2007. Trophic cascades triggered by overfishing reveal possible mechanisms of ecosystem regime shifts. Proceedings of the National Academy of Sciences 104: 10518-10523.

105. Kaiser, M.J., J.S. Collie, S.J. Hall, S. Jennings, and I.R. Poiner. 2003. Impacts of fishing gear on marine benthic habitats. pp. 197–217 *In*: M.

Sinclair and G. Valdimarsson (eds.). Responsible Fisheries in the Marine Ecosystem. FAO, Rome.

106. Roberts, 2007.

107. Kaiser et al., 2003.

108. Johannes, R.E. 1981. Words of the lagoon: fishing and marine lore in the Palau district of Micronesia. University of California Press. Berkeley.

109. McClellan, K. 2008. Coral degradation through destructive fishing practices. The Encyclopedia of Earth. Source: http://www.eoearth.org/article/Coral_degradation_through_destructive_fishing_practices?topic=49513. Accessed on 9-14-2012.

110. FAO. 2005. World inventory of fisheries. Destructive fishing practices. Issues Fact Sheets. Text by S.M. Garcia. In: FAO Fisheries and Aquaculture Department website. Source: http://www.fao.org/fishery/topic/12353/en. Accessed on 10-14-2012.

111. The Coral Reef Alliance. 2008. Exploitive fishing. Coral Reef Alliance Issue Briefs. Source:http://www.coral.org/node/130. Accessed on 10-13-2012.

112. Fox, H.E. and R.L. Caldwell. 2006. Recovery from blast fishing on coral reefs: a tale of two scales. Ecological Applications 16(5):1631-1635.

113. Russell, B. (Grouper & Wrasse Specialist Group) 2004. *Cheilinus undulatus*. In: IUCN 2012. IUCN Red List of Threatened Species. Version 2012.1. Source:

http://www.iucnredlist.org/search/details.php/4592/all. Accessed on 5-25-2012.

114. Pet-Soede, L. and M. Erdmann. 2008. An overview and comparison of destructive fishing practices in Indonesia. SPC Live Reef Fish Information Bulletin 4:28-36.

115. Rabanal, H.R. 1988. The history of aquaculture. FAO Fisheries and Aquaculture Department. Source: http://www.fao.org/docrep/field/009/ag158e/AG158E02.htm. Accessed on 8-30-2012.

116. Sumaila et al., 2010.

117. Bert, T.M. 2007. Environmentally responsible aquaculture--a work in progress. Pp. 1-31 In: T.M. Bert (ed.), Ecological and genetic implications of aquaculture activities. Springer, Dordrecht, The Netherlands.

118. Bert, 2007.

119. Kahru, M., B.G. Mitchell, A. Diaz, and M. Miura. 2004. MODIS detects a devastating algal bloom in Paracas Bay, Peru. Eos, Transactions American Geophysical Union 85(45):465-472.

120. Grescoe, 2008.

121. Clover, 2006.

122. FAO. 2012. The state of the world's fisheries and aquaculture 2012. FAO Fisheries and Aquaculture Department. Source: http://www.fao.org/docrep/016/i2727e/i2727e.pdf. Accessed on 11-19-2012.

123. About CCAMLR. Commission for the Conservation of Antarctic Marine Living Resources. Source: http://www.ccamlr.org/en/organisation/about-ccamlr. Accessed on 10/21/2012.

124. Moran, S. 2012. Team tracks a food supply at the end of the world. New York Times. Source: http://www.nytimes.com/2012/03/13/science/tracking-antarctic-krill-as-more-is-harvested-for-omega-3-pills.html. Accessed on 10-21-2012.

125. Dekker, W. 2008. Coming to grips with the eel stock slip-sliding away. Pp. 335-355 *In*: M.G. Schlecter, N.J. Leonard, and W.W. Taylor (eds.). International governance of fisheries ecosystems: learning from the past, finding solutions for the future. American Fisheries Society Symposium 58. Bethesda, Maryland.

126. Russell, I.C. and I.C.E. Potter. 2003. Implications of the precautionary approach for the management of the European eel, *Anguilla anguilla*. Fisheries Management and Ecology 10:395-401.

127. Leber, K.M. 2004. Summary of case studies on the effectiveness of stocking aquacultured fishes and invertebrates to replenish and enhance coastal fisheries. Pp. 203-213 *In*: D.M. Bartley and K.M. Leber (eds.). Marine ranching. FAO Fisheries Technical Paper. No. 429. FAO, Rome.

128. McEachron, L.W., C.E. McCarty, and R.R. Vega. 1993. Successful enhancement of the Texas red drum (*Sciaenops ocellatus*) population. Pp. 53-6 *In*: Interactions between cultured species and naturally occurring species in the environment. Proceedings of the 22nd US Japan Aquaculture Panel Symposium.

129. Ottolenghi, F., C. Silvestri, P. Giordano, A. Lovatelli, and M.B. New. 2004. Capture-based aquaculture: the fattening of eels, groupers, tunas and yellowtails. Rome, FAO.

130. Ellis, 2008.

131. Ottolenghi, F. 2008. Capture-based aquaculture of bluefin tuna. Pp. 169-182 In: A. Lovatelli and P.F. Holthus (eds). Capture-based aquaculture. Global overview. FAO Fisheries Technical Paper. No. 508. Rome, FAO.

132. Ellis, 2008.

133. Bert, 2007.

134. Hannah, R.W. and S.A. Jones. 2007. Effectiveness of bycatch reduction devices (BRDs) in the ocean shrimp (*Pandalus jordani*) trawl fishery. Fisheries research, 85(1): 217-225.

135. Hannah, R.W. and Jones, S.A. (2003). Measuring the height of the fishing line and its effect on shrimp catch and bycatch in an ocean shrimp (*Pandalus jordani*) trawl. Fisheries Research, 60(2): 427-438.

136. Lichatowich, J. 1999. Salmon without rivers: a history of the Pacific salmon crisis. Island Press. Washington, D.C.

137. Araki, H., B. Cooper, and M.S. Blouin. 2007. Genetic effects of captive breeding cause a rapid, cumulative fitness decline in the wild. Science 318(5847):100-103.

138. Hilborn, R. 2006. Salmon-farming impacts on wild salmon. Proceedings of the National Academy of Sciences, 103(42):15277.

139. Volpe, J.P., E.B. Taylor, D.W. Rimmer, and B.W. Glickman. 2000. Evidence of natural reproduction of aquaculture-escaped Atlantic salmon in a coastal British Columbia river. Conservation Biology 14: 899-903.

140. Costello, M.J. (2009). How sea lice from salmon farms may cause wild salmonid declines in Europe and North America and be a threat to fishes elsewhere. Proceedings of the Royal Society B: Biological Sciences 276(1672):3385-3394.

141. Hites, R.A., J.A. Foran, D.O. Carpenter, M.C. Hamilton, B.A. Knuth, and S.J. Schwager, 2004. Global assessment of organic contaminants in farmed salmon. Science 303(5655): 226-229.

142. Closed containment. Solutions. Farmed and dangerous website. Source: http://www.farmedanddangerous.org/solutions/closed-containment/. Accessed on 11/3/2012.

143. Upton, H.F. and E.H. Buck. Aug. 9, 2010. Open ocean aquaculture. CRS Report for Congress. Source: http://www.respecttheocean.org/linked/congressional_doc.pdf. Accessed on 9-3-2012.

144. Puerto Rico at the cutting edge of offshore aquaculture. Sept. 23, 2002. NOAA's Office of Oceanic and Atmospheric Research. Archive of

Spotlight Feature Articles. Source: http://www.oar.noaa.gov/spotlite/archive/spot_snapperfarm.html. Accessed on 9-1-2012.

145. Puro, R. June 6, 2011. Startup profile: an open ocean farm where fish never see the same water twice. Seedstock. Source: http://seedstock.com/2011/06/06/an-open-ocean-farm-where-fish-never-see-the-same-water-twice/. Accessed on 9-3-2-12.

146. Buttner, J.K. and G. Karr. 2009. East meets West: Hawai'i, a lesson for aquaculture development in the United States. Part I: the early days. World Aquaculture 40(4):41-44+.

147. Helsley, C.E. 2000. Hawai'i open ocean aquaculture demonstration program. In Proceedings of the 28th US-Japan Natural Resources Aquaculture Panel Joint Meeting on Spawning and Maturation of Aquaculture Species, Technical Report No. 28, pp. 15-22.

148. Hukilau Foods website. Source: http://www.hukilaufoods.com/about_us. Accessed on 8-31-2012.

149. Hukilau Foods files for bankruptcy. Nov. 3, 2010. Honolulu Star Advertiser. Source: http://www.staradvertiser.com/business/20101103_Hukilau_Foods_files_for_bankruptcy.html?id=106591479. Accessed on 8-31-2012.

150. Hukilau Foods plans $1.5M moi hatchery. May 27, 2011. Pacific Business News. Source: http://www.bizjournals.com/pacific/print-edition/2011/05/27/hukilau-foods-plans-15m-moi-hatchery.html?page=all. Accessed on 8-31-2012.

151. Dennis, K. 2005. Hawaii's 2nd open-ocean farm launches Kona Kampachi. Center for Tropical and Subtropical Aquaculture Regional Notes 16(3 & 4):1,8.

152. Simpson, S. 2011. The blue food revolution. Scientific American 304:54-61.

153. Kona Kampachi successful harvest in federal waters off Hawaii. March 01, 2012. Hawaii Free Press. Source: http://www.hawaiifreepress.com/ArticlesMain/tabid/56/articleType/ArticleView/articleId/6236/Kona-Kampachi-Successful-Harvest-in-Federal-Waters-off-Hawaii.aspx. Accessed on 8-30-2012.

154. Blue Ocean completes acquisition of Kona Blue mariculture assets. June 4, 2012. Blue Ocean Mariculture, News and events. Source: http://www.bofish.com/2012/06/. Accessed on 8-30-2012.

155. Kona Blue. Feb. 10, 2012. Guy Harvey Magazine. Source: http://www.guyharveymagazine.com/topics/sustainable-seafood/kona-blue. Accessed on 9-3-2012.

156. Fish farm leaves Culebra for Panama. 2009. The Free Library. Source: http://www.thefreelibrary.com/Fish+farm+leaves+Culebra+for+Panama.-a0200729990. Accessed on 9-3-2012.

157. Marine Aquaculture Task Force. 2007. Sustainable marine aquaculture: fulfilling the promise; managing the risks. Marine Aquaculture Task Force, Takoma Park, MD.

158. Sen. Judd Gregg celebrates nation's first commercial offshore mussel farm. Oct. 11, 2007. Atlantic Marine Aquaculture Center news release. Source: http://ooa.unh.edu/news/releases/2007-10_gregg/mussels_release.html. Accessed on 9-3-2012.

159. Shellfish aquaculture. 2007. Atlantic Marine Aquaculture Center website. Source: http://ooa.unh.edu/shellfish/shellfish_about.html. Accessed on 9-3-2012.

160. Upton and Buck, 2010.

161. Bardach, J.E. 1997. Fish as food and the case for aquaculture. Pp. 1-14 *In*: J.E. Bardach (Ed.), Sustainable aquaculture. John Wiley & Sons, Inc. New York.

162. Bert, 2007.

163. Marine Aquaculture Task Force, 2007.

164. GAO. 2011. Seafood safety: FDA needs to improve oversight of imported seafood and better leverage limited resources. United States Government Accountability Office Report to Congressional Requesters. GAO-11-286. Source: http://www.gao.gov/assets/320/317734.pdf. Accessed on 11-10-2012.

165. Import Alert 16-24. 2012. Detention without physical examination of aquaculture seafood products due to unapproved drugs. US FDA. Source: http://www.accessdata.fda.gov/cms_ia/importalert_27.html. Accessed on 11-16-2012.

166. Love, D.C., S. Rodman, R.A. Neff, and K.E. Nachman. 2011. Veterinary drug residues in seafood inspected by the European Union, United States, Canada, and Japan from 2000 to 2009. Environmental Science & Technology 45:7232-7240.

167. Lumpkin, M.M. 2007. Safety of Chinese imports. FDA Congressional Testimony, July 18, 2007. Source: http://www.fda.gov/NewsEvents/Testimony/ucm110728.htm. Accessed on 11-16-2012.

168. FDA News Release. 2007. FDA detains imports of farm-raised Chinese seafood: products have repeatedly contained potentially harmful residues. Source: http://www.fda.gov/NewsEvents/Newsroom/Press Announcements/2007/ucm108941.htm. Accessed on 11-12-2012.

169. Editorial. 2009. Melamine and food safety in China. Lancet 373(9661):353.

170. Zhang, J., D.L. Mauzerall, T. Zhu, S. Liang, M. Ezzati, and J.V. Remais. 2010. Environmental health in China: progress towards clean air and safe water. Lancet 375: 1110–1119.

171. Guo, J.Y., E.Y. Zeng, F.C. Wu, X.Z. Meng, B.X. Mai, and X.J. Luo. 2007. Organochlorine pesticides in seafood products from southern China and health risk assessment. Environmental Toxicology and Chemistry 26(6):1109–1115.

172. FAO, 2012.

173. Watson, R. and D. Pauly. 2001. Systematic distortions in world fisheries catch trends. Nature 414:534-536.

174. Worm et al., 2006.

175. Kurlansky, 1997.

176. Harris, M. 1998. Lament for an ocean: The collapse of the Atlantic cod fishery, a true crime story. McClelland & Stewart Inc. Toronto.

177. Clover, 2006.

178. The Prince's Charities International Sustainability Unit. 2012. Towards global sustainable fisheries: the opportunity for transition. Source: http://www.pcfisu.org/wp-content/uploads/2012/01/ISUMarine programme-towards-global-sustainable-fisheries.pdf. Accessed on 12-2-2012.

179. Bellido, J.M., M.B. Santos, M.G. Pennino, X. Valeiras, and G.J. Pierce. 2011. Fishery discards and bycatch: solutions for an ecosystem approach to fisheries management? Hydrobiologia 670:317–333.

180. Charles, A. and L. Wilson. 2009. Human dimensions of marine protected areas. ICES Journal of Marine Science 66:6–15.

181. Sumaila et al., 2010.

182. Pauly, D., V. Christensen, S. Guenette, T.J. Pitcher, U.R. Sumaila, C.J. Walters, R. Watson, and D. Zeller. 2002. Towards sustainability in world fisheries. Nature 418:689-695.

183. Radovich, J. 1982. The collapse of the California sardine fishery: what have we learned? CalCOFI Report 23:56-78.

184. Park, J.W. and T.M.J. Lin. 2005. Surimi: manufacturing and evaluation. Pp. 33-106 *In*: J.W. Park (Ed.), Surimi and surimi seafood, second edition. CRC Press. Boca Raton, FL.

185. Park, J.W. 2005. Surimi seafood: products, market, and manufacturing. Pp. 375-433 *In*: J.W. Park (Ed.), Surimi and surimi seafood, second edition. CRC Press. Boca Raton, FL.

186. LeSann, A. 1998. A livelihood from fishing. Intermediate Technology Publications, London.

187. Datta, S. 2009. Intrinsic & extrinsic parameters of fish spoilage. Scribd. Source: http://www.scribd.com/doc/19877519/Intrinsic-Extrinsic-Parameters-of-Fish-Spoilage. Accessed on 9-26-2012.

188. Kurlansky, 1997.

189. Roche, J. and M. McHutchinson (eds.). 1998. First fish first people: salmon tales of the North Pacific rim. University of Washington Press. Seattle.

Index

anchoveta 102, 127, 129
anchovy 67-68, 118
antibiotic residues 99, 124, 126
aquaculture 5, 61, 97-109, 111-128
   extensive 71, 97, 120
   intensive 97-99, 126

benefits of seafood diet 12-22
   brain development 17, 32-34
   cancer 16-18
   during pregnancy 20, 32-34
   heart disease 16, 18-21, 80
   nutrition 12, 18-19, 22
benthic habitats 63-64, 71, 92-93
biomagnification 24-26, 29
bivalves 36-37, 51-54, 71, 119, 128
bluefin tuna 12, 31, 81-82, 88, 91, 107
bycatch 85-91, 103, 109-110, 121, 131
   dolphins 85-86, 89
   Norwegian policy 90
   seabirds 85-88
   sharks 86
   turtles 85-86, 88-89
bycatch reduction devices 85, 89-90
   circle-shaped hooks 88
   turtle excluding devices 89, 109
   tori lines 88

canneries 25, 135
capture fisheries 61, 66, 91-96, 98-99, 102, 106-107, 121, 125, 127, 133
capture-based aquaculture 107

carbon dioxide emissions 10
carp 102, 120-121, 124, 128
catfish 26, 57, 71, 121, 124
China 78-79, 94-97, 101, 123-126, 128
climate, influence of ocean on 3-5, 131
coastal marine ecosystems 5-6, 108, 131
cobia 116
cod 7, 21, 58, 69-70, 75, 130, 135, 142
collapse of fisheries 11, 82, 104-105, 108, 129
   cod 7, 130
   sardines 135
conservation of marine resources 65, 104, 130
contamination 21, 24, 27, 30, 34, 49, 99, 102, 124-125
coral reef ecosystems 5, 10, 43, 65, 92-96
crab 10, 15, 36-37, 42-43, 50, 54, 63, 69, 71, 107, 130, 138-139, 145-147, 168
   Dungeness 71, 145
   stone 71

dams 111
dead zones 8-10, 93
deep-sea habitats 74, 96
Deepwater Horizon 9, 82
destructive fishing practices 92-96, 131
   blast fishing 94-95, 126
   bottom trawls 64, 74, 92, 110
   dredges 63-64, 71, 92

Ecological Food for Thought on Seafood

muro ami 94
poisoning 95-96
diatoms 37, 42
dinoflagellates 36-37, 41, 47
dioxins 27, 30, 102
disease 14, 16, 18-21, 52-53, 80, 97-99, 107-109, 113-114, 120, 122-123, 126
domoic acid 37, 42-43

ecological impact of capture fisheries 91-96
ecosystem-based fisheries management 130
eels 44, 101, 104-105, 107
efficient utilization of fish 120, 137-143
effluent 8, 101, 125
El Niño 4, 127, 129
endangered species 79, 82, 85, 87
environmental impact 73, 109, 119, 122
escapement 112, 119
exclusive economic zones (EEZs) 58, 61, 84

filter feeders 36, 51, 67, 71, 97, 121, 128
fish meal 66, 68, 98, 101-104, 109, 113, 118
fish oil 15, 17-22, 67-68, 70, 101-103, 118, 154
fish poisoning 37, 43-46, 117
  ciguatera 37, 42-45, 117
  fugu 37, 45-46
fisheries 4, 7, 11-12, 55, 57-96, 98-99, 101-104, 106-107, 109, 119, 121, 125, 127-135, 140-142
  artisanal 58-59, 61, 93-95, 140
  commercial 12, 119, 127-129
  industrial 61, 66-68, 101-103, 106-107, 113, 125

fishing advisories 26-27, 30, 44-45
fishing gear 59-60, 63-64, 74, 85, 91-93, 96, 131
  bottom trawl 64, 74, 92, 110
  dredge 63-64, 71, 92
  hook and line 25, 63-64, 154
  longline 63, 70, 87-88, 121
  purse seine 63, 86, 94
  traps 57, 59, 61, 63, 70, 90, 110, 121
fishing, history of 57-62
fishing technology 57, 74-75
food advisories 26-27, 30-32, 34, 45
food chain 4, 12, 23-31, 35-36, 43-44, 66-67, 91, 98, 103-104, 106, 114
Fukushima 8
furans 30, 124

ghost fishing 89-91, 96
government fishing subsidies 83, 91, 99, 132-133
Greenland Eskimos 21
grouper 44, 64-65, 96, 106-107, 141
guidelines 12, 20, 26-27, 30-33
  dining 12, 71-73
  FDA, for cooking oysters 53
  FDA and EPA for consumption of fish and shellfish 26-27, 30-31
Gulf of Mexico 5, 9, 51, 65, 71, 82, 88-89, 107
Gulf Stream 3, 105

habitat complexity 92-93
halibut 15, 44, 70, 75
harmful algal blooms (aka red tides) 36-37, 42, 46-47, 101
hatcheries 10-11, 111-112
Hawaiian yellowtail 117-119

heavy metals 16, 21, 23-28, 34, 99, 126
herring 20, 66-67, 161
high grading 85, 91
hooks 25, 57, 59, 63-64, 86-88
hotline, shellfish 50
hurricanes 5, 9, 96

illegal, unreported, unregulated (IUU) fishing 76, 78, 84, 87-88, 90, 105, 133
impact of aquaculture on wild fisheries 101-106
   feed 101-104
   seed 104-106
impact of cage pens on wild salmon 99-100, 104, 106, 112-113
import alert 123-124
imported seafood 123-126
   antibiotic residues 124, 126
   contamination 124-125
   drug residues 123, 126
   safety of 123, 125-126
indigenous conservation 142-143

Japan 8, 41, 45-46, 66, 70, 81-82, 88, 94, 98, 104-107, 117, 123, 136, 138, 142

kamaboko 139-140
kelp beds 5, 15-16
krill 103-104

Law of the Sea 58

mackerel 13, 15, 20, 25-26, 32, 44, 68-69, 101, 140
mahi-mahi (aka dorado) 45, 69
mangroves 5-6, 108-109
marine protected areas (MPAs) 131-132
marine ranching 106

menhaden 22, 66-68, 101
mercury 21, 23-29, 31-35, 72, 117, 125, 154
moi 116-117
mussels 14, 36, 38-42, 50-52, 54, 57, 71, 93, 107, 119-122, 128, 159, 164-166
myths, shellfish 51-54

net pens 97, 99, 106, 111-112, 115
nuclear wastes 8

ocean acidification 10
ocean circulation 3-4, 9, 23, 64-65
ocean dumping 7-8, 78, 85, 90-91, 131
omega-3 fish oils/fatty acids 12, 18-22, 34, 67-68, 70, 72, 154
open-sea fish farming 115-119, 121-122
orange roughy 74-75, 77
Oregon 10, 25, 34, 43, 110, 115, 145, 167
overcapacity 83-84, 129
overfishing 7, 72-74, 81-84, 91, 95, 102, 131, 140-141
oysters 14-15, 36, 41, 49-54, 63, 71, 97, 107, 121, 128, 159

Pacific hake 139
Patagonian toothfish (aka Chilean sea bass) 62, 74-76, 136
PCBs 8, 21, 27-30, 72, 102, 113, 117
pesticides 21, 27, 95, 102, 109, 113, 126
pirate fishing 75-76, 84, 91, 133
plankton 3, 19, 24, 35-37, 47, 66-67, 103, 121
puffer fish (aka fugu) 45-46

recipes 145-168
   albacore 146, 153, 160

Ecological Food for Thought on Seafood   187

crab 145-147, 168
mussels 159, 164-166
oysters 159
salmon 146, 148-152, 155-158, 168
shrimp, northern pink 167
reef fishes 45, 94-96
rock hoppers 92

sablefish (aka black cod) 70-71, 75
salmon 13, 15, 20, 23, 26-27, 43, 66, 69-70, 97-99, 111-115, 140, 142-143
sand lance 101, 103
sardines 13-15, 65-68, 135, 137
scallops 14, 36, 41, 54, 63-64, 71, 92, 138
sea lice 113, 115
seafood production, state of the world's 127-129
seafood safety 22, 30, 41, 44-46, 49-51, 123, 126
seahorses 79-80
seaweed (aka sea vegetables) 15-16, 96
shark-fin trade 78-79
sharks 74, 77-79, 86
shellfish 9-10, 12, 15, 26, 29, 36-43, 46-47, 49-54, 57, 61, 71, 97, 99, 115, 120-121, 123, 126, 136, 139, 143
shellfish poisoning 37-43, 47
 amnesiac (ASP) 37, 42-43
 diarrhetic (DSP) 37, 41-42
 history of 37-42
 neurotoxic (NSP) 37, 42
 paralytic (PSP) 37-42, 47
shrimp 6, 15, 24, 26, 36, 54, 63, 80, 88-89, 92, 101, 107-110, 124-125, 130, 167

cold-water 110, 130
farmed 100, 108-110, 124-125
wild-caught 88-89, 109-110
spoilage 35, 49-50, 59, 140
squid 29, 36, 62, 71, 103, 136
stock enhancement 83, 106, 117
surimi 69, 138-142
sushi 12, 16, 72-73, 81, 105, 107, 117
sustainable aquaculture 117, 120-122
sustainable fisheries 55, 62-73, 132-133
sustainable seafood choices 66-73
swordfish 15, 24-26, 32, 45, 88
synthetic fibers 60, 89-90

threatened species 77-78, 80, 82, 89, 96
tilapia 71, 102, 121, 123-124
toxins 16, 26, 35-52
traditional Chinese medicine (TCM) 79-80
tsunami 5, 8, 81
tuna 12, 15, 20, 23-26, 30-32, 45, 57, 68, 81-82, 86, 88, 91, 107-108

underutilized species 134-136

Vancouver, Capt. George 38-40
Vibrio 51-53

wahoo 45, 68-69
walleye pollock 26, 44, 69, 139-140
wild versus farmed comparison 98-100, 109-114
 fish 111-114
 shrimp 109-110

Acknowledgements

Much of the information presented here has been gleaned from a variety of published and Internet sources. I have made a concerted effort to attribute specific facts to their original sources. But widely-published, general principles are presented without specific attribution. I acknowledge the huge contribution that reading scientific and popular marine literature has made to this project. And I thank all the authors for both informing me, and entertaining me along the way.

I would like to thank Carol Roe for asking the question that began this journey. And I gratefully acknowledge the years of support and encouragement provided by Bori Olla, along the way. Special thanks are due to David Ehrenfeld, who reviewed an earlier version of this book, and Becky Anderson for her assistance editing the final draft.

Having access to the electronic, physical, and intellectual resources of Oregon State University's Guin Library at the Hatfield Marine Science Center greatly facilitated my search for knowledge. Libraries provide multiple portals into the world of information.

CPSIA information can be obtained at www.ICGtesting.com
Printed in the USA
BVOW03s1346200913

331653BV00004B/9/P